TRISOMY G / NORMAL MOSAICISM

W0050512

E. S. SACHS M.D.

TRISOMY G / NORMAL MOSAICISM
A cytological and clinical investigation

1971
H. E. STENFERT KROESE N.V. / LEIDEN

The present investigation was carried out in the department of pathology of the University Hospital of Leiden, The Netherlands, under supervision of Dr. A. Schaberg, professor of pathology, and Dr. H. H. van Gelderen, professor of paediatrics.

ISBN-13: 978-90-207-0311-5   e-ISBN-13: 978-94-010-2956-8

DOI: 10.1007/ 978-94-010-2956-8

Copyright 1971 by H. E. Stenfert Kroese N.V.

LEIDEN / THE NETHERLANDS

CONTENTS

Acknowledgements

Advice was given by Dr. M. van der Ploeg, department of histochemistry, and Dr. W. Th. Daems and Drs. P. Brederoo of the department of electron microscopy, and Miss M. Hoogerwerf, department of medical statistics, University of Leiden. The patients were studied at the *Dr. Mr. Willem van den Bergh Stichting,* Noordwijk (director: Dr. T. Pouwels), the *Johannesstichting,* Nieuwveen (director: Dr. J. Hoeing), the department of anthropobiometrics of the University of Amsterdam (head: Dr. J. B. Bijlsma), and the out-patient paediatrics department of the University Hospital, Leiden. The technical assistance was given by Miss M. van den Berge. The Figures were prepared by Mr. G. Flippo. The manuscript was typed by Mrs. A. S. de Haas. The investigation was made possible by financial support from the foundation *De Drie Lichten,* and the *Dr. Mr. Willem van den Bergh Stichting,* Noordwijk.

I wish to express my thanks to all for their kind co-operation.

INTRODUCTION

The normal human chromosome number is 46, 22 pairs of autosomes and two X chromosomes or an X and a Y chromosome. The G group contains the two smallest pairs of chromosomes, numbered 21 and 22. The occurrence of an extra chromosome in this group is called trisomy G and this phenomenon is present in all cells in patients with Down's syndrome.

Trisomy G/normal mosaic patients who have an extra G chromosome in a certain percentage of their cells, have been described by various authors. This chromosomal pattern has been detected mainly in patients with signs of Down's syndrome and has also been reported in nine mentally retarded patients in whom Down's syndrome was not suspected (van Gelderen et al. 1967). In addition trisomy G/normal mosaicism can occur in phenotypically normal persons, as has been the case in several parents of children with Down's syndrome.

Cytogenetic studies carried out in a group of patients from institutions for mentally retarded persons and from out-patient departments, yielded ten trisomy G/normal mosaic patients with varying percentages of trisomic cells in skin and blood. Phenotypically, this group varied from atypical to the typical appearance of Down's syndrome.

These ten patients were examined more extensively to evaluate the relationship between the degree of mosaicism and various phenotypical characteristics of Down's syndrome. A group of Down's syndrome patients and normal controls were included in the investigations.

These investigations comprised clinical examinations, studies of dermatoglyphs, and the radiological aspects of the pelvic bones. Leukocyte alkaline phosphatase was determined both cytochemically and biochemically. Electron-microscopical studies were carried out to detect eventual morphological changes in the granules of neutrophil granulocytes related to enzyme levels.

1

A. THE LITERATURE

I. CHROMOSOME STUDIES

Lejeune et al. (1959) first described the presence of an extra G chromosome in Down's syndrome patients. Up to now, it has not been possible to distinguish with certainty between chromosomes 21 and 22. In the case of patients with clinical Down's syndrome the chromosome is usually numbered 21. A number of patients have been described with an extra G chromosome without the clinical stigmata of Down's syndrome, in which cases it has been suggested that this might be a chromosome 22. A review of these patients has been given by Chaudhuri et al. (1968).

In Down's syndrome patients with translocation the extra chromosome is attached to another chromosome, usually one of the D-group or G-group. Both types of translocation occur more frequently sporadically than familially (Soltan et al. 1964, Sergovich et al. 1964).

Trisomy G mosaicism, where the extra G chromosome occurs only in some of the cells, was first observed by Clarke et al. in 1961. Since then, many patients with this chromosomal aberration have been described.

A survey of the literature is given in Tables I, II, and III, in which the patients have been divided clinically into three groups: those with clinical signs of Down's syndrome, those with doubtful clinical signs, and those lacking clinical signs of Down's syndrome, who are phenotypically either normal or abnormal. The patients of the first group (Table I) i.e. those with clinical signs of Down's syndrome, were detected by chance during studies of large groups of patients with Down's syndrome, because they were born to young mothers or their chromosomes were analysed when they were hospitalized for other reasons. The trisomy G mosaic patients with doubtful clinical signs (Table II) were analysed on the basis of the dubious picture. The trisomy G mosaic patients of the last group (Table III, A and B), i.e. those without clinical signs of Down's syndrome, were mostly detected when they were karyotyped because they had given birth to one or more mongoloid children (Table III A). The nine patients previously described by van Gelderen et al. (1967) (Table III B) belonged

2

Table I. Data on trisomy G/normal mosaic patients
with clinical signs of Down's syndrome, as reported in the literature

Author	Year	No. of patients	Detected mosaics	Sex	Age (yr)	% Trisomic cells		Maternal age
						Leuk.	Fibrobl.	
Brøgger	1966	90	1					
				M	2	20	99	24
Chitham	1964	105	3					
and Mac Iver				M	21	4	80	29
				M	20	50	80	45
				M	15	57	90	38
Chu Ch'ang Ning	1966	25	2					
and Chung Lien-Yun				F	5	63	—	37
				M	9/12	64	—	37
Edgren et al.	1966	73	2					
				F	15		73	24
				M	25	61	46	32
Finley et al.	1966	82	4					
				M	6	20-55	—	24
				F	5	12	—	—
				M	4	15-40	—	23
				F	10	38-64	—	21
Greyerz-Gloor et al.	1969	272	8					
				4 M	—	12.5	—	four
				and	—	20.5		older
				4 F	—	40.0	—	than
					—	49.0	—	30
					—	57.1	—	four
					—	58.8	—	younger
					—	61.1	—	than
					—	86.0	—	30
Gustavson	1961	1	1					
and Ek				M	12	—	38	24
Hayashi	1963	83	2					
				M	8	52	—	37
				M	3	43	—	17
Huang et al.	1967	77	1					
				F	12	75	—	32

3

Table I *(Continued)*

Author	Year	No. of patients	Detected mosaics	Sex	Age (yr)	% Trisomic cells Leuk.	% Trisomic cells Fibrobl.	Maternal age
Kohn et al.	1970	8	8	M	12	4	—	22
and		(Previously		M	1	30	56	22
Taysi et al.		identified		M	2/12	74	88.6	23
		as mosaics)		M	6	74	—	20
				F	8½	51	95.3	38
				M	21	3.5	38	36
				M	28	62	94	16
				M	42	30	—	—
Marks et al.	1967	1	1					
				M	6	83	—	23
Mikkelsen	1967	100	2					
				M	1	44	30	25
				M	2	84	82	16
Nichols et al.	1962	1	1					
				F	2/12	17	67	19
Richards et al.	1965	225	6					
				—	33	75	100	25
				—	33	6	13	33
				—	6	49	100	22
				—	28	13	60	36
				—	41	7	56	44
				—	17	23	76	31
Ridler et al.	1965	1	1					
				M	52	1	72	45
Taylor	1968	11	11					
		(Previously		F	2	45	82	—
		identified		F	1½	58	94	—
		as mosaics)		F	2	1	27	—
				F	3½	3	67	—
				F	18	18	100	—
				F	1	8	2	—

4

Table I *(Continued)*

Author	Year	No. of patients	Detected mosaics	Sex	Age (yr)	% Trisomic cells Leuk.	% Trisomic cells Fibrobl.	Maternal age
Taylor	1968	11	11					
		(Previously		M	1½	95	—	—
		identified		F	1	98	—	—
		as mosaics)		M	4 days	72	—	—
				F	33	5	6	—
				F	37	1	3	—
Tonomura et al.	1966	127	3					
				M	2/12	74	—	32
				F	3	58	—	35
				F	—	92	—	—
Tsuboi and Inouye	1968	1	1	F	5	73	—	38
Valencia et al.	1963	1	1	M	7/12	30	—	38
Warkany et al.	1964	58	2	M	22	60	—	25
				M (Table II)				
Weiss and Wolf	1968	2		M	2 weeks	67	—	38
				M	9	68	—	32
Zellweger and Abbo	1963	1	1	M	3	68	—	40
Zellweger et al.	1966	1	1	M	5	7	61	22

5

Table II. Data on trisomy G/normal mosaics with doubtful signs
of Down's syndrome, as reported in the literature

Author	Year	No. of patients	Detected mosaics	Sex	Age (yr)	% Trisomic cells Leuk.	% Trisomic cells Fibrobl.	Maternal age
Aula et al.	1961	1	1	F	1	32 (bone marrow)	—	25
Blank et al.	1962	1	1	F	31	13	—	40
Chaudhuri and Chaudhuri	1965	1	1	M	4	59	—	22
Clarke et al.	1963	1	1	F	2	14	32-38	26
Fitzgerald and Lycette	1961	1	1	M	51	53	—	38
Giraud et al.	1963	1	1	F	17	—	34	39
Giraud et al.	1965	1	1	M	6	—	23	34
Hamerton et al.	1965	173	6	—	31	42	—	29
				—	27	20	12	26
				—	30	18	—	35
				—	30	73	—	32
				—	38	4	—	28
				—	55	81	—	45
Hirsch et al.	1967	2	2	F	3½	50	—	25
				F	3	74	—	28

Table II (Continued)

Author	Year	No. of patients	Detected mosaics	Sex	Age (yr)	% Trisomic cells		Maternal age
						Leuk.	Fibrobl.	
Lindsten et al.	1962	1	1					
				F	2	61	27	38
Mauer and Noe	1964	1	1					
				M	3½	72	—	38
Petit and Gallez	1969	1	1					
				F	5½	20	—	31
Pfeiffer	1966	312	10					
				F	24	10	—	27
				M	3	40	—	28
				F	6	38	—	38
				—	1½	42	—	42
				No data are given for the other 6 patients				
Reinwein et al.	1966	67	4					
				F	3½	35	60	26
				F	1 month	—	87	24
				M	7/12	83	100	42
				F	1 month	16	—	24
Tsuboi and Inouye	1968	1	1					
				M	6	40	—	33
Warkany et al.	1964	1	1					
				M	23	0	23	23
Walker and Ising	1969	2	2					
				F	6	15	—	—
				M	—	23	—	—

7

Table III. Data on trisomy G/normal mosaics without clinical signs of Down's syndrome, as reported in the literature

Author	Year	No. of patients	Detected mosaics	Sex	Age (yr)	% Trisomic cells Leuk.	% Trisomic cells Fibrobl.	Maternal age
A. Phenotypically normal patients.								
Aarskog	1969	2	1	F	25	5	—	39
Ferrier	1964	2	2	F	25	25	20	27
				M	26	21	—	24
Smith et al.	1962	1	1	F	19	27	75	39
Verresen et al.	1964	1	1	F	32	10	—	34
Waxman and Arakaki	1966	1	1	F	—	0	22	—
Weinstein and Warkany	1963	1	1	F	17	16	18	39

B. Mentally retarded patients with multiple congenital anomalies.

Van Gelderen et al.	1967	38	9	M	—	36	—	32
				F	—	6.5	—	32
				F	—	9	—	30
				M	—	9	—	43
				F	—	7	—	39
				F	—	13	—	27
				F	—	15	—	37
				F	—	30	—	36
				F	—	40	30	24

to a series with multiple congenital anomalies, growth retardation, and otherwise unexplained mental deficiency.

According to Penrose (1964), mosaicism is considered to be doubtful when the percentage of normal or trisomic cells in a patient is lower than nine. The mosaicism of some of the patients described in the literature becomes questionable when this criterion is applied.

Hamerton et al. (1965) described one patient with only two trisomic cells in a count of 50 cells and one patient with 25 trisomic cells and one normal cell in a count of 30 cells. Taylor (1968) described eleven trisomy G/normal mosaic patients in five of whom the percentage of cells with either 46 or 47 chromosomes was below 9. One of the patients described by Aarskog (1969) has only 5 per cent trisomic leukocytes, but has given birth to two boys with Down's syndrome, which confirms her trisomy G mosaicism.

A trisomy G/normal mosaic patient can originate from either a trisomic or a normal zygote (Bartalos and Baramki 1967). When a trisomic zygote, caused by non-disjunction in the mother, loses the extra chromosome by anaphase lagging in some of the cell lines originating from this zygote, a trisomy G/normal mosaicism occurs in the child. A normal zygote may produce, by non-disjunction, trisomic and monosomic cell lines, the latter being non-viable. When one or more of the trisomic cell lines lose the extra chromosome by anaphase lagging, mosaicism is the result. Non-

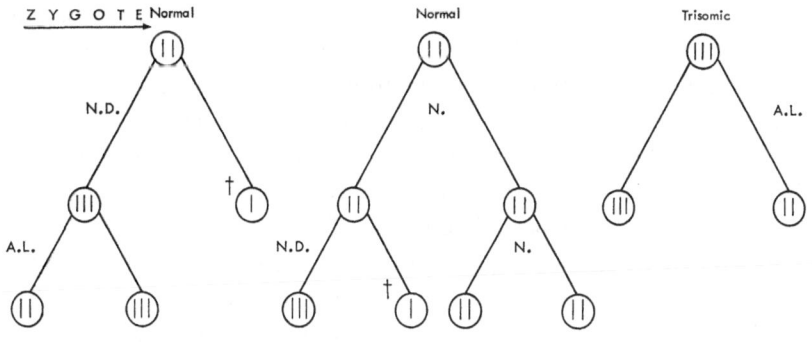

N. = normal division.　　N.D. = non-disjunction.　　A.L. = anaphase lagging.

Fig. 1. Possible origins of trisomy G/normal mosaicism.

9

disjunction of a normal zygote at a later division produces trisomic cell lines in addition to the existing normal cell lines. Fig. 1 illustrates these possibilities. Fig. 2 shows karyograms of a regular and a mosaic Down's syndrome patient.

Penrose (1963, 1964) has stated that the occurrence of trisomic cells is about twice as frequent in fibroblasts as in leukocytes in individual mosaic patients. Ford (1969) ascribes this difference to varying proliferation rates of normal and abnormal cells, the effect of a slower differentiation of trisomic cells becoming more pronounced in the rapidly dividing leukocytes. Richards (1969) has given an extensive survey of cell distribution in 230 mosaic patients based on data obtained by sending questionnaires to investigators in many countries. The author has pointed out that there is a strong bias in the selection of patients for skin cultures, probably favouring cases with a low proportion of trisomic cells in the blood. In his opinion, the mean percentage of 71.3 trisomic cells in the skin cultures of the trisomy G mosaics might be as high as 85 or even higher if the skin cultures had been carried out in all mosaic mongols. A lower percentage of trisomic cells in fibroblasts is found only in the patients of Mikkelsen (1967), Hamerton et al. (1965), Lindsten et al. (1962), van Gelderen et al. (1967), Edgren et al. (1966), and Taylor (1968). More or less equal percentages of trisomic leukocytes and fibroblasts were found in patients of Mikkelsen (1967), Ferrier (1964), Weinstein and Warkany (1963), and Taylor (1968).

A rapid cell selection of small lymphocyte stem cells in young mosaic mongols has been observed by Taylor in two studies (1968, 1970). During the first two or three years of life, either normal or trisomic cells may be selected. Richards (1969) data confirmed this evidence by showing an increase of normal cells in some patients before the age of one year. Although as a rule an equilibrium of the two cell-lines seems to be reached between the age of one and two years, two adults with shifts of proportions of cell lines have been described by Taysi et al. (1970).

Besides 46-47 mosaicism, caused by the extra G chromosome additional chromosome changes have been reported by various authors. Triple cell line mosaicism (46-47-48) caused by cells trisomic and tetrasomic for a G chromosome, are described by Tonomura and Takehiko (1964), Fitzgerald and Lycette (1961), and Mauer and Noe (1964). Tsuboi and Inouye's (1968) patient showed a 46-47 mosaicism caused by trisomy G and an extra telocentric G chromosome in the cells with 48 chromosomes. The patient of Valencia et al. (1963) had 4-7 acrocentrics in 39 per cent of bone marrow cells. A D-G translocation occurred in one of Ferrier's (1964) mosaic patients, whilst the mother described by Waxman and Arakaki (1966) has normal blood cells and a 22 per cent G-G translocation mosaicism in skin cells.

Mosaicism for the G chromosome and one other chromosome was found by Edgren et al. (1966) in a patient with 45-46-47 mosaicism caused by cells with 45-XO, 46-XY, 47-XY G+ chromosomes. The patient of Gustavson and Ek (1961) also had a triple form (46-47-48) caused by mosaicism for a G and an F chromosome, and Marks et al. (1967) saw a patient with an alternative trisomy G and trisomy 18 in cells with 47 chromosomes. A case of double autosomal trisomy with mosaicism for chromosomes 18 and 21 is described by Yu-Feng et al. (1965). Weiss and Wolf (1968) described two brothers with trisomy G mosaicism, which was thought to be the result of mitotic non-disjunction secondary to a familial balanced C-G translocation present in the mother.

II. CLINICAL INVESTIGATIONS

1. Physical signs and intelligence level

The various authors use different criteria for the clinical diagnosis of Down's syndrome, which makes it difficult to evaluate the occurrence of clinical signs of this syndrome in mosaic patients with a typical or atypical form.

Pfeiffer (1966) described ten mosaic patients found in a series of 312 patients investigated either for genetic reasons, because they were born to young mothers, or for diagnostic reasons, because of a doubtful clinical diagnosis of Down's syndrome. According to the author, eight of these ten patients were clinically not mongoloid, but in our opinion the photographs of four of them show definite signs of Down's syndrome. This seems logical, since he calls his 312 patients 'mongoliens'.

A correlation between percentages of trisomic cells and mongoloid stigmata has been suggested by Reinwein et al. (1966), who investigated 67 children of younger mothers and of older mothers with incomplete Down's syndrome. Four mosaic patients were detected. This relationship seems to be contradicted by the existence of phenotypically

11

normal patients (Table III A) and by the patients with normal intelligence described by Clarke et al. (1963), Hayashi et al. (1962), and Lindsten et al. (1962).

Rosecrans (1968) investigated a possible correlation between the intelligence of twenty mosaic patients and their percentage of abnormal cells. He chose these twenty patients out of thirty-one reported in the literature, because these were the only cases in which he considered the method of measuring the intelligence to be acceptable. The mean IQ of this sample was 65, which is far above the mean of mongoloid populations. He concluded that the percentage of abnormal cells in either skin or blood has some predictive value with respect to intellectual development in mosaicism. This conclusion is based on a positive correlation found between either percentages of abnormal cells in skin cultures or the highest percentage of abnormal cells in skin or blood on the one hand and the level of intelligence on the other.

Von Greyerz-Gloor et al. have found no significant difference in clinical symptoms between eight mosaic patients detected in a group of 272 mongoloid patients and the patients of this group with a regular trisomy G.

Kohn et al. (1970) pointed out the difficulty involved in attempting to correlate the proportion of trisomic cells, which may be variable, and the IQ test results, which are known to decrease with increasing chronological age in Down's syndrome (Koch et al. 1963). They found no apparent correlation between the percentage of trisomic cells and the IQ in eight trisomy G/normal mosaic patients.

Chaudhuri and Chaudhuri (1965) concluded that it is impossible to establish a definite relationship between the proportions of various cell lines and physical features or level of intellectual attainment, and this seems to be confirmed by the contradictory statements of the other authors.

The clinical findings of the group of patients described by van Gelderen et al. (1967) consisted of retarded growth, multiple congenital anomalies, and mental deficiency. The patients were selected for cytogenetic studies because there was no known aetiology for their symptoms. Five of these patients will be described in the following chapters.

Phenotypical signs of Down's syndrome, though mentioned more or less extensively in published cases of trisomy G mosaicism, have not yet been analysed systematically. This would have been not very useful in any case,

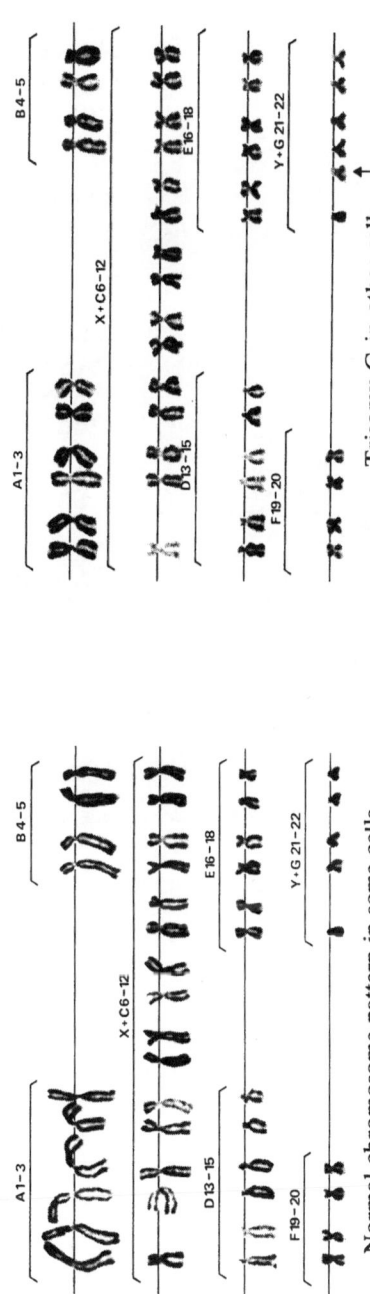

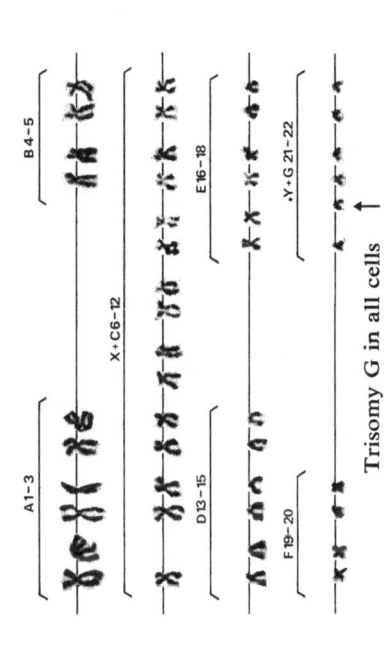

Fig. 2. Karyograms of a trisomy G/normal mosaic patient and a patient with Down's syndrome.

because of the selection applied: virtually all of the patients described in the literature were selected on the basis of clinical signs of Down's syndrome, the only exceptions being the clinically normal parents of children with Down's syndrome (Table III A) and the patients described by van Gelderen et al. (1967). The variety of criteria chosen by various authors for the clinical diagnosis also interferes with the analysis of the criteria themselves.

2. Dermatoglyphs

Dermatoglyphs have been shown to be highly characteristic in patients with Down's syndrome (Ford Walker 1958). The dermatoglyphs of thirteen trisomy G mosaic patients described in the literature have been discussed by Penrose (1965).

Digital patterns showed the characteristic ulnar loop on all ten digits in 4.2 per cent of the controls, 23 per cent of the mosaic patients and 32 per cent of the Down's syndrome patients. The palmar triradii showed an average deviation in mosaic mongols of 78 per cent toward the mongol average, which is highly significant statistically. Tibial arches and open fields, caused by the absence of a hallucal triradius, are found in 0.5 per cent of the controls and in 47 per cent of the Down's syndrome patients on the right and left soles (Ford Walker 1958). Of the trisomy G/normal mosaic patients described by Penrose (1965), 50 per cent showed no hallucal triradius. The percentages of occurrence of these dermatoglyphic patterns in trisomy G/normal patients were therefore intermediate between those found for controls and Down's syndrome patients in the case of ulnar loops and palmar triradius and equal to that found in Down's syndrome patients in the case of hallucal patterns consisting of open field and tibial arch.

Dermatoglyphic data of additional trisomy G/normal mosaic patients are shown in Table IV. Ten ulnar loops and a high palmar triradius have been found in a percentage equal to that in Down's syndrome patients, and three tibial arches have been seen in two patients, giving an incidence intermediate between that of controls and Down's syndrome patients. In thirteen patients with more than 50 per cent trisomic cells in either blood or skin, 22 out of 26 hands, or 85 per cent, show a distal axial triradius (t″). In six patients with less than 50 per cent trisomic cells, only one out of twelve hands showed a distal axial triradius (t″), i.e. 8.5 per cent.

13

LongTable IV. *Dermatoglyphs of 21 trisomy G/normal mosaic patients, as reported in the literature*

Patient	1	2	3	4	5	6	7	8	9	10	11	12	13	14	15	16	17	18	19	20	21
Trisomic cells																					
Leukocytes	10	19	.	25	.	35	50	7	30	59	7	64	67	68	1	74	74	83	74	62	51
Fibroblasts	.	.	23	20	34	38	.	56	56	.	61	.	.	.	72	.	.	89	94	95	
Fingertips																					
Left digits 1	A	U	.	U	A	U	.	U	U	.	.	W	.	W	U	.	.	U	U	U	U
2	U	U	.	U	W	U	.	U	U	.	.	U	.	U	U	.	.	U	U	U	U
3	R	U	.	U	U	U	.	U	U	.	.	U	.	U	U	.	.	U	U	U	U
4	U	U	.	U	W	U	.	U	U	.	.	U	.	R	U	.	.	U	U	U	U
5	U	U	.	U	U	U	4	U	U	.	7	U	.	W	U	7	.	U	.	U	U
							X				X					X					
							U				U					U					
Right digits 1	U	U	.	U	U	U	+	U	U	.	+	W	.	W	U	+	.	U	U	U	U
2	U	U	.	U	U	U	6	W	A	.	3	U	.	U	U	3	.	U	U	U	U
3	U	U	R	R	U	U	X	U	U	.	X	U	.	U	U	X	.	U	U	U	U
4	R	U	.	U	W	W	W	U	U	.	W	U	.	W	W	W	.	U	U	U	U
5	U	U	.	U	W	U	.	U	U	.	.	U	.	W	U	.	.	U	U	U	U

Pattern 3rd interdigit. area

Palmar triradius

Hallucal pattern

	1	2	3	4	5	6	7	8	9	10	11	12	13	14			
Pattern 3rd interdigit. area L	+	.	.	+	+	.	.	.	+	.	.	.	+	.	.	.	.
Pattern 3rd interdigit. area R	+	.	.	+	+	.	.	.	+	.	.	.	−	+	.	.	.
Palmar triradius L	t	t′	.	t	t	t	.	.	t″	t	t″	t″	.	t″	t	t′	t′
Palmar triradius R	t′	t″	.	t	t	t	.	.	t″	t	t″	t″	.	t″	t′	t′	t″
Hallucal pattern L	L.tib	.	.	LLD	.	W	LD	LLD	W	.	.	A.tib	.	A.tib	SLD	LLD	LLD OF
Hallucal pattern R	L.tib	.	.	LLD	.	W	LD	LLD	W	.	.	L.tib	.	A.tib	LLD	LLD	LLD OF
Author	1	2	3	4	5	6	7	8	9	10	11	12	13	7	6	14	6 6 6

Authors
1. Verresen et al. 1964
2. Petit et al. 1969
3. Giraud et al. 1965
1. Ferrier 1964
5. Giraud et al. 1963
6. Kohn et al. 1970
7. Hirsch et al. 1967
8. Richards et al. 1965
9. Chaudhuri et al. 1965
10. Zellweger et al. 1966
11. Chu Ch'ang Ning et al. 1966
12. Weiss et al. 1968
13. Ridler et al. 1965
14. Marks et al. 1967

U   = Ulnar loop
R   = Radial loop
A   = Arch
W   = Whorl
LLD   = Large distal loop
SLD   = Small distal loop
OF   = Open field
L.tib = Loop tibialis
A.tib = Arch tibialis

The one patient of Richards et al. (1965) and one of Ridler et al. (1965) with a low percentage of trisomic cells in leukocytes but a high percentage in fibroblasts, both have a high palmar triradius. This suggests that dermatoglyphic abnormalities are positively correlated with the degree of mosaicism. The phenotypically normal patient described by Verresen et al. (1964) has a radial loop on the 4th finger of the right hand, which when present is highly suggestive of Down's syndrome.

3. Pelvic measurements

No studies have been published, to the best of our knowledge, on pelvic measurements in trisomy G mosaics, though such data are known to be of great diagnostic value in patients with Down's syndrome (Caffey and Ross 1956).

III. MATERNAL AGES

With respect to maternal ages, it must be kept in mind that patients with Down's syndrome are often selected for karyotyping because of the youth of the mothers. This holds for the cases described by four out of eleven authors (Mikkelsen 1967, Pfeiffer 1966, Reinwein et al. 1966, Warkany et al. 1964), and led to the young maternal age of 16 of their 37 trisomy G/normal mosaic patients. The very young maternal ages of the four patients of Finley et al. (1966) possibly also introduced a bias in his series. In Pfeiffer's (1966) series the maternal ages of five of his ten mosaic mongols lay above 30 years, the other five being between 29 and 27 years.

A detailed analysis of the maternal ages of trisomy G mosaic patients has been given by Richards (1969), who argues that trisomy G/normal mosaics originating from trisomic zygotes would have the same mean maternal age at birth as Down's syndrome patients, and trisomy G/normal mosaics originating from normal zygotes the same maternal age at birth as normal children. The drop in the mean maternal age at birth from 33.3 years for trisomic patients to the 31.9 years of his 47 trisomy G/normal mosaic patients could, according to this hypothesis, be explained if 20 per cent of the mosaic patients originated from normal zygotes.

Maternal ages of trisomy G/normal mosaic patients are therefore certainly elevated; the smaller elevation of the maternal age compared with that of Down's syndrome patients might be caused by differences in the origin and the selection of the mosaic patients.

IV. LEUKOCYTES

The mean level of leukocyte alkaline phosphatase shows an even greater elevation in mosaic patients than in Down's syndrome patients (van Gelderen et al. 1967). For leukocytes of trisomy G/normal mosaics, mean nuclear lobe counts lie intermediate between those for Down's syndrome patients and controls. No other studies of leukocyte alkaline phosphatase activity in trisomy G mosaicism have been reported.

The occurrence of chronic myeloid leukaemia had been described in the families of the patients of Waxman and Arakaki (1966) and Verresen et al. (1964).

The literature on Down's syndrome relevant to present investigations will be discussed in the appropriate places.

B. THE PRESENT INVESTIGATIONS

I. INTRODUCTION AND SELECTION OF PATIENTS

The literature on trisomy G mosaicism reviewed in the preceding chapter does not clarify the influence on phenotypical characteristics of the extra G chromosome in a percentage of the cells. A group of trisomy G/normal mosaic patients was therefore studied, together with a group of trisomic patients and a group of normal controls, to collect more information about this relationship. Special attention was given to the occurrence of the phenotypical characteristics of Down's syndrome patients. The investigations included the following subjects.

Chromosomes were studied in blood and skin cultures. Clinical examinations included measurements of body height and head circumference and evaluation of the characteristic features of Down's syndrome. Dermatoglyphs of the trisomy G/normal mosaics, their parents, and the Down's syndrome patients were analysed. Pelvic measurements of the trisomy G/normal mosaic patients were compared with those of Down's syndrome patients and matching controls. The maternal ages and birth weights of the trisomy G/normal mosaic patients were also analysed. Leukocyte alkaline phosphatase levels and lobe counts were determined for all three groups. The leukocytes were investigated electron-microscopically to detect possible morphological changes related to enzymatic changes and thus localize the enzyme activity.

The trisomy G/normal mosaic material comprised ten patients, five of whom belonged to the group described by van Gelderen et al. (1967). As mentioned above, these patients were detected during karyotyping of 38 mentally retardet patients with multiple congenital anomalies and retarded growth in an attempt to demonstrate the etiology of their symptoms. Therefore, these five patients were not selected because they showed some or all of the features of Down's syndrome or had mongoloid offspring, as is the case for all the other patients described in the literature. On the contrary, possible mosaics with typical features of Down's syndrome may have been missed. It seemed worthwhile to collect additional data concerning these patients. Three of our mosaic patients were investigated on the basis of a dubious diagnosis of Down's syndrome, and the chromosomes of two patients were studied because they were the first mongoloid child of a young mother.

Only mosaic patients with more than 9 per cent of either normal or trisomic cells in the leukocyte or fibroblast cultures, for whom karyotyping demonstrated the extra chromosome to belong to the G group and the cells with 46 chromosomes showed a normal karyotype, were classified as trisomy G/normal mosaics. Until a definite distinction can be made between chromosomes 21 and 22, we suggest that patients with an extra G chromosome in some of their cells should be classified as trisomy G/normal mosaics.

Thirty Down's syndrome patients, all of whom were karyotyped, were investigated to study the incidence of the investigated signs in Down's syndrome patients. Twenty-five of these patients are in an institution for the mentally retarded (the *Dr. Mr. Willem van den Bergh Stichting* in Noordwijk), the remaining five are out-patients of the Department of Paediatrics of the Leiden University Hospital.

The signs of Down's syndrome were also studied in a group of 80 normal children from 1 to 5 years of age registered at a welfare clinic, since several of these signs occasionally also occur in normal persons. The relationship between age and the appearance or disappearance of certain physical signs will be discussed.

II. CHROMOSOME STUDIES

Chromosome studies were carried out on leukocytes in the trisomy G/normal mosaics as well as in all Down's syndrome patients, and additional fibroblast cultures were made for eight trisomy G/normal mosaics. We tried to count at least 50 cells from each culture, and when this was not possible, to repeat the culture.

1. Methods

Leukocytes: Twelve ml of venous blood is obtained from fasting patients in a syringe containing 1 ml heparin (400 U per ml). This is divided into two portions, each of which is added to a test tube containing

1 ml sterilized dextran (5 per cent, mol.weight 200,000). If the blood has to be transported, this should be done in a container provided with ice. Sedimentation at 37° C should take place for 1 to 1½ hours with the tubes standing at a 45° angle. Plasma and leukocytes are then pipetted off, taking care not to disturb the erythrocytes. Sterile containers (20 ml) are filled with 3 ml medium 199 (Morgan et al. 1950) to which penicillin (100 U per ml) has been added. The colour of the medium should be orange-red; if it is purple, 5 per cent CO_2 should be passed through it until the colour is right. Then, 1.5 ml plasma containing leukocytes and 0.25 ml phytohaemagglutinin solution is added to each container. The phyto solution is prepared by adding 5 ml phytobuffer to frozen phyto M powder, both from Difco Laboratories, Detroit, Michigan 48201, U.S.A. For one culture, 4 to 5 containers are needed. The cells are incubated at 37° C for 72 hours, and the containers should be shaken gently every 24 hours.

After 72 hours, 0.25 ml 0.04 per cent colcemide is added to each container and the containers are left to stand for 2 hours at 37° C. The cells are then resuspended in the containers, and the total content of the containers is divided over 2 test tubes, which are rotated for 7 minutes at 80 x g. The supernatant fluid is pipetted off, and 6 ml hypotonic solution, consisting of 1 part Hanks solution and 3 parts distilled water at 37° C is added to each test tube. The cells are then resuspended, left for 20 minutes at 37° C, and again gently suspended. The tubes are now rotated for 7 minutes at 80 x g, after which the supernatant fluid is pipetted off and 4 ml of a fresh mixture of 3 parts methanol and 1 part glacial acetic acid added to the tubes. This must be done very gently, letting the first drops slide separately along the side of the test tubes. The cells are left in the fixative for 30 minutes, after which they are collected by rotation for 5 minutes at 110 x g. Most of the fixative is then carefully aspirated off, 0.5 ml of fresh fixative is added, and the cells are resuspended. Slides have been placed in cold water at 4° C previously. A few drops of the cell suspension are brought onto each slide after the sides and lower surface have been blotted with filter paper to remove excess water. The slides are then dried above a flame and shaken to dry them quickly without allowing them to become hot, stained with orcein, and kept overnight at 4° C. The orcein solution is made by adding 2 gr of orcein pur to 50 ml 90 per cent lactic acid and 50 ml 45 per cent acetic acid, and is filtered before use. The slides are differentiated with 45 per cent acetic acid the next morning, put through a graded alcohol series (twice in 96 per cent, twice in 100 per cent) and then in xylol for 20 minutes. They are covered with a thin layer of malinol and mounted. In some cases the Difco micromethod was used for leukocyte cultures (Difco Laboratories, Detroit, Michigan 48201, U.S.A.).

Fibroblasts: Skin-biopsies are obtained from the lateral side of the upper arm or leg, the skin first being carefully disinfected with 70 per cent alcohol. No anaesthetic is used, since the specially designed rotating drill needs only be applied for one second. Two or three biopsy samples with a diameter of ± 3 mm are taken for each culture and brought immediately into medium 199. They are then cut, lying in a few drops of the medium, under a dissection microscope with cataract knives. The pieces are washed separately in fresh medium 199 and placed together in a drop of medium 199 on a slide resting on ice under sterile cover.

These small pieces are then divided over 4 or 5 small Carrel flasks, to each of which is added 1 ml of medium consisting of: 60 per cent medium 199, 20 per cent human AB serum, 20 per cent chicken embryonic juice, and penicillin, 100 U per ml. The flasks are closed with rubber stoppers and incubated at 37° C.

They should then be left for about a week, after which time the medium should be refreshed when it turns yellow. When the bottom of the flasks is covered with fibroblasts, the cells are trypsinized. The medium is aspirated off and replaced by 0.2 per cent trypsin for 3 to 5 minutes at 37° C. When the cells have become detached from the bottom of the bottle, they are suspended in the trypsin and the suspension is brought into a test tube filed with Ca^{++} and Mg^{++} free Hanks solution (to prevent clustering of the cells). After rotating for 7 minutes at 80-100 x g, the supernatant fluid is discarded and the cells are resuspended in 1 ml medium consisting of 70 per cent medium 199, 20 per cent human AB serum, 10 per cent chicken embryonic juice, and penicillin, 100 U per ml. The suspension is then brought into a 2.5 ml Carrel flask. The medium is refreshed when it turns yellow. When there is sufficient cell growth, trypsinizing is repeated and the cell suspension is brought into a 10 ml flask. Cells were harvested on 3 or 4 occasions between 3 and 6 weeks after the setting up of the cultures. No definite duration can be given for the cell growth of the patient material because, unlike that of controls, it tends to differ widely.

The cell harvesting is carried out as follows: 1 ml of a 0.04 per cent colcemid solution diluted 100 times with Ca++ and Mg++ free Hanks solution is added to a 10 ml culture flask, which is then incubated for 6 hours at 37° C. After 6 hours the medium is discarded and replaced by a layer of 0.2 per cent trypsin. The flask is held at 37° C. When after 3 to 5 minutes the cells have become detached from the bottom of the bottle, they are resuspended and the suspension is added to a test tube with 5 ml Ca++ and Mg++ free Hanks solution and rotated for 7 minutes at 80-100 x g. The supernatant fluid is then pipetted off, hypotonic solution is added, and the culture is treated further in the same way as the leukocyte cultures after the hypotonic solution has been added.

2. Results

The results of the chromosome analyses of the trisomy G/normal mosaic patients are shown in Table V. Patients 1, 2, 3, and 6 have a higher percentage of trisomic cells in their fibroblasts than in their leukocytes. Patient 3 has almost exclusively trisomic fibroblasts and the percentage of his trisomic leukocytes is low, i.e. 92 per cent and 13 per cent, respectively. Patients 8 and 10 have roughly equal percentages of trisomic fibroblasts and trisomic leukocytes and patients 4 and 5 a lower percentage of trisomic fibroblasts as compared with the leukocytes. All the Down's syndrome patients had an extra G chromosome in their cells; two of the 30 patients showed a D-G translocation, in the others there was trisomy.

A remarkable finding is the high percentage of aneuploidy in most patients in addition to the trisomy G. Cell cultures in our laboratory as a rule have between 5 and 10 per cent aneuploid cells. We have found a higher percentage (about 15) of aneuploidy other than trisomy G, in the cultures of regular and mosaic trisomy G patients and in cell cultures of patients with other cerebral defects. This phenomenon could perhaps be explained by a general tendency toward irregular cell division caused either by non-disjunction or by anaphase lagging, which is also responsible for the occurrence of the trisomy G or the trisomy G/normal mosaicism.

Table V. Results of chromosome studies of the trisomy G/normal mosaic patients

					Leukocyte cultures						Fibroblast cultures						
					% of cells with chromosome count of:								% of cells with chromosome count of:				
Pat. No.	sex	age	n*	n**	<45	45	46	47	>47	age	n*	n**	<45	45	46	47	>47
1.	F	21	100	10	8	11	73	8	—	25	58	—	—	—	78	22	—
		22	50	10	8	14	72	6	—								
		22	50	2	10	12	70	8	—								
2.	M	14	50	7	2	12	74	12	—	17	68	—	—	4	81	15	—
		14	50	9	4	12	78	6	—								
3.	M	15	47	3	6	2	79	13	—	17	27	2	—	4	4	92	—
4.	F	17	100	6	—	3	84	13	—	21	73	3	—	6	90	3	1
		18	50	6	2	2	82	14	—								
5.	F	21	100	14	3	14	68	15	—	24	35	2	—	6	83	11	—
6.	M	18	60	12	8	7	52	30	3	22	51	—	2	2	43	51	2
7.*	F	12	30	.	3	3	34	60	—	—	—	—	—	—	—	—	—
8.	M	7/12	12	—	8	—	17	75	—	1½	45	5	—	2	20	78	—
		9/12	63	13	5	5	24	66	—								
		1	21	5	10	10	—	80	—								
9.*	F	9/12	40	.	—	2.5	17.5	80	—	—	—	—	—	—	—	—	—
10.	M	5	67	7	7	—	9	84	—	5	32	3	3	—	9	85	3

n* = number of cells counted
n** = number of cells investigated

* Chromosome studies were carried out at the Department of Anthropobiometrics of the University of Amsterdam.

III. CLINICAL EXAMINATIONS

In this chapter the body height, the cranial circumference, and the ratio upper/lower segment of the trisomy G/normal mosaic patients will be discussed, as well as the individual characteristics of Down's syndrome in these patients. Ten physical signs were considered to be of diagnostic value in Down's syndrome on the basis of their divergent incidence in Down's syndrome patients and normal controls of the same population groups. These ten signs were: epicanthus, Brushfield's spots, abnormal ears, furrowed tongue, abnormal teeth, short neck, curved digit V, plantar furrow, cutis marmorata, and scant body hair.

Certain physical signs must be evaluated according to the age of the patient, because one of the most characteristic signs of Down's syndrome, the furrowed tongue, does not occur in newborns but appears after the second year (Engler 1949). Another characteristic sign of Down's syndrome, epicanthus, is also present in many normal children, in whom it disappears gradually in the first years. For this reason, we believe that the incidence of these features should be viewed separately in children and adults. The incidence of the investigated parameters must also be known for Down's syndrome patients and normal controls, to evaluate their incidence in trisomy G/normal mosaicism correctly. The results of the various investigations in the trisomy G/normal mosaic patients will therefore be compared with those of Down's syndrome patients and controls.

The results of the analysis of the dermatoglyphs of the trisomy G mosaics, their parents, and the Down's syndrome patients will be discussed.

Since pelvic measurements are known to be of diagnostic value in young children with Down's syndrome, we also analysed these data for the trisomy G/normal mosaic patients and groups of Down's syndrome patients and controls matched for sex and age.

1. Body measurements

Body measurements of trisomy G/normal mosaic patients are of interest since deviations have been demonstrated by various authors in Down's syndrome patients.

24

An extensive study of growth and sexual maturation in children with cerebral defects has been made by Dooren (1967) in a series of 65 patients with Down's syndrome, 28 girls and 37 boys. Only one girl had a height above the 10th percentile curve; the heights of three boys lay only temporarily above this curve during puberty. These results are in agreement with those of Mosier et al. (1965), where the 253 patients with Down's syndrome were the shortest group of the mental defectives. Swaak (1967) reported values above the 10th percentile length curve for boys and girls with Down's syndrome in 19 and 13 out of 137 and 98 measurements, respectively, for 125 children under 3 years of age. This is not in contradiction with the above-mentioned data, because dwarfism may be less pronounced in very young patients with Down's syndrome.

Dooren (1967) and Mosier et al. (1965) also demonstrated that Down's syndrome patients and patients with hypothyroidism are the only groups of mental defectives showing a relative shortening of the lower extremities. The head circumferences of patients with Down's syndrome have been shown to lie below normal values (Benda 1969, Mosier et al. 1965).

On the basis of these findings we therefore measured body height, symphysis-heel height, and head circumference in our trisomy G/normal mosaics and Down's syndrome patients and compared these values with those found in normal boys and girls. The symphysis-heel height was measured with the patient in the standing position. This value was subtracted from the total height and the ratio

$$\frac{\text{total height} - \text{symphysis-heel height}}{\text{symphysis-heel height}} = \frac{\text{upper segment}}{\text{lower segment}}$$

was calculated. The results were as follows.

a. Body height
The crown-heel heights of 17 male Down's syndrome patients and 5 male trisomy G/normal mosaic patients are shown in Fig. 3. All heights of the Down's syndrome patients and the trisomy G/normal mosaic patients are on or below the 10th percentile curve for normal boys of the same age in The Netherlands (de Wijn and de Haas 1960).

The crown-heel heights of 13 female Down's syndrome patients and 5 female trisomy G/normal mosaic patients are shown in Fig. 4. Both groups, with the exception of one Down's syndrome girl, also showed heights on or below the 10th percentile curve for normal girls of the same age in The Netherlands (de Wijn and de Haas 1960).

25

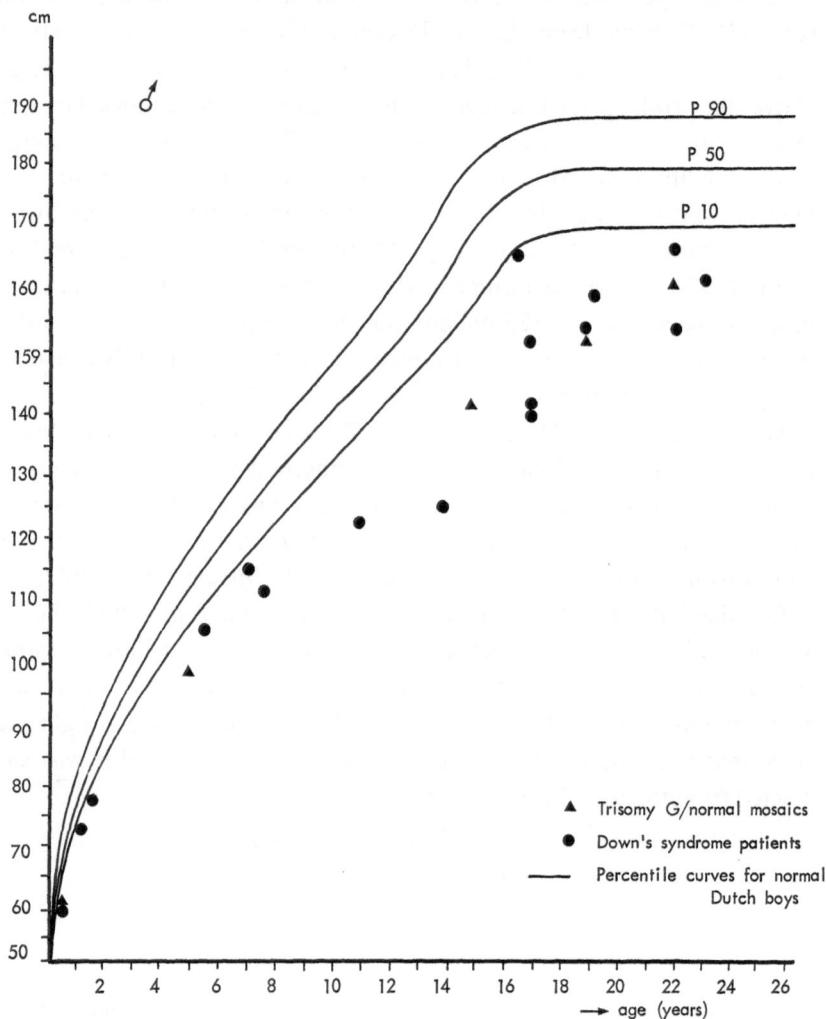

Fig. 3. Body heights of male patients.

b. Upper/lower segment ratio
The shortening of the lower extremities characteristic of Down's syndrome patients also occurred in our trisomy G/normal mosaic patients, as can be seen in Figs. 5 and 6. These ratios were calculated for patients older than 2 years, because under this age the values increase rapidly.

26

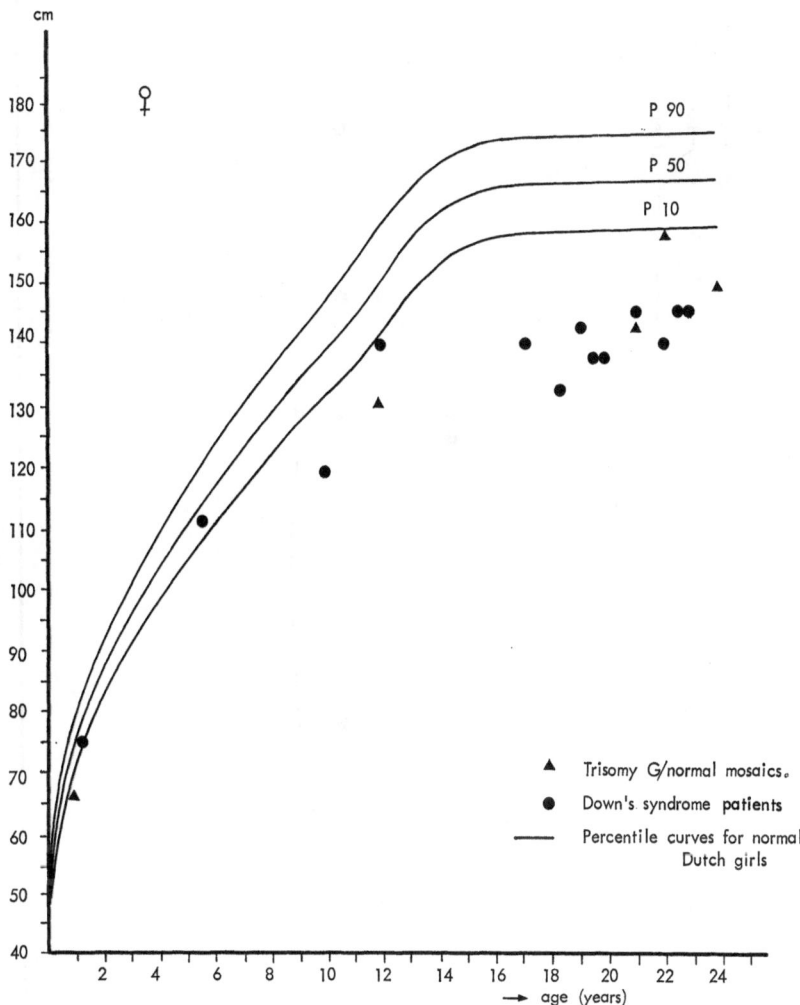

Fig. 4. Body heights of female patients.

c. Head circumference

The values found in our cases of Down's syndrome and trisomy G/normal mosaic patients, shown in Figs. 7 and 8, are in agreement with those found by Mosier et al. (1965). All the patients had small head circumferences.

It is clear from the results that the values for total height, upper/lower segment ratio, and head circumference show no differences between the groups of Down's syndrome patients and trisomy G/normal mosaics. The

27

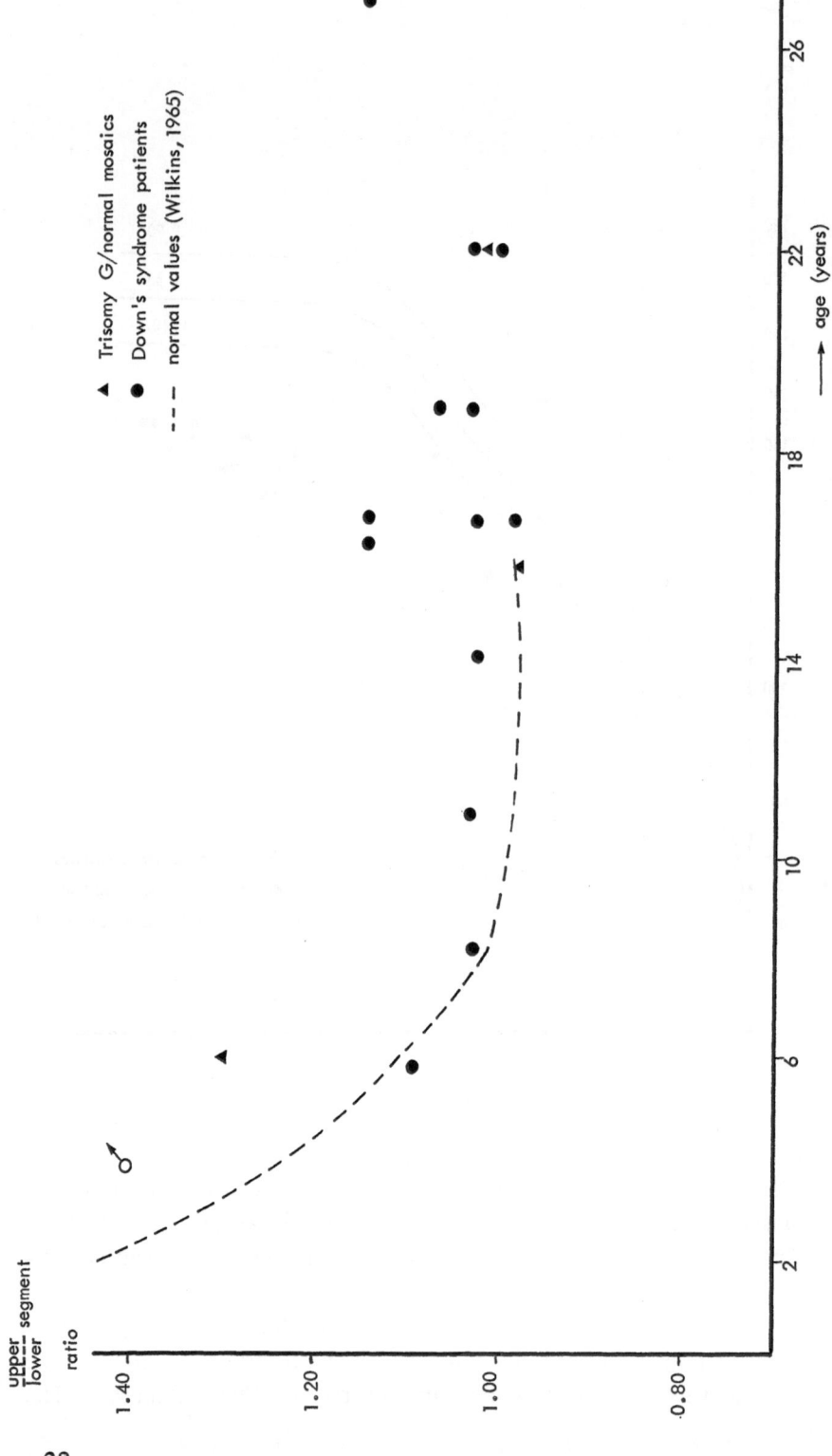

Fig. 5. Upper/lower segment ratios of male patients.

▲ Trisomy G/normal mosaics
● Down's syndrome patients
--- normal values (Wilkins, 1965)

upper segment
lower
ratio

1.40

1.20

1.00

0.80

2   6   10   14   18   22   26

⟶ age (years)

28

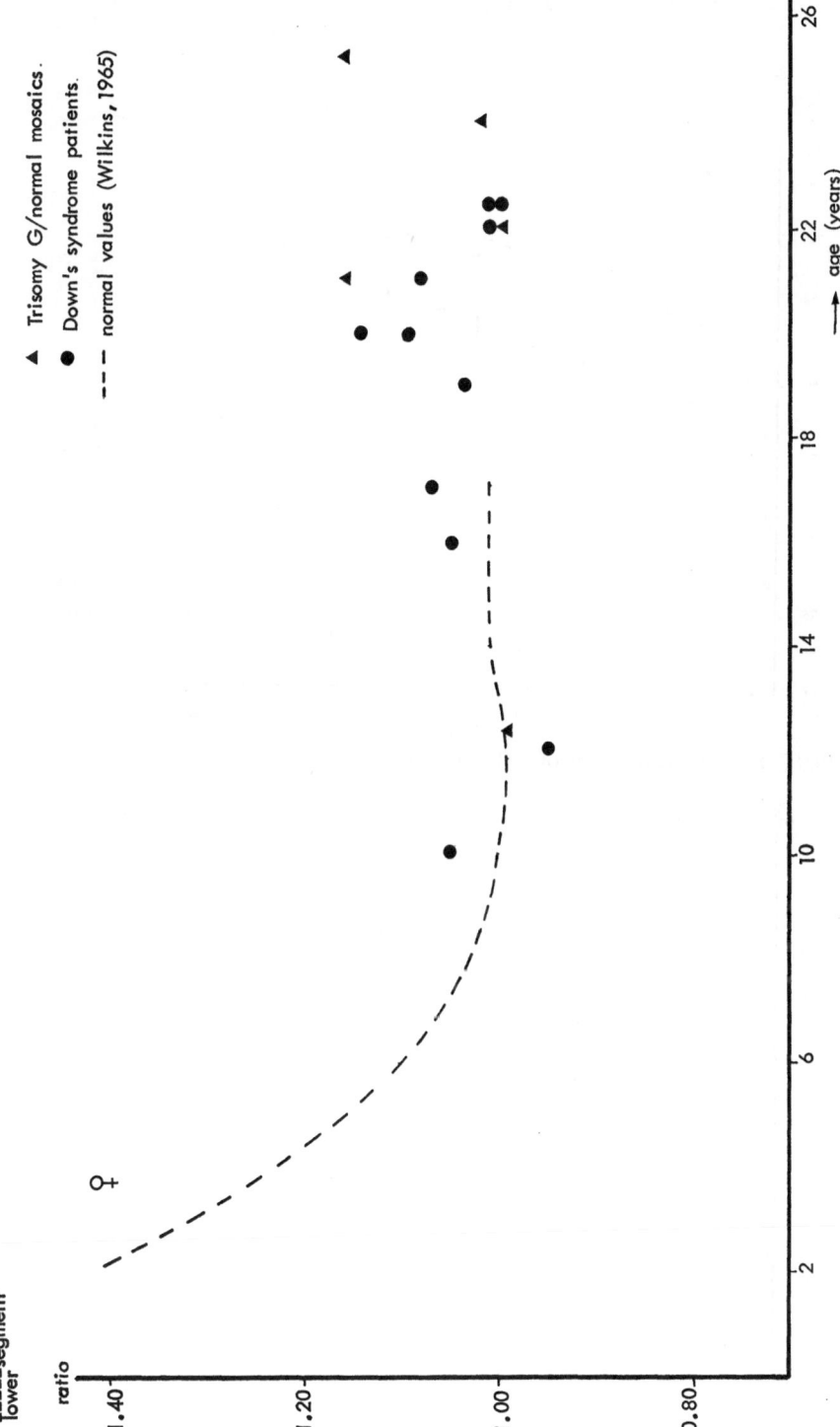

upper segment
lower segment

ratio

▲ Trisomy G/normal mosaics.
● Down's syndrome patients.
--- normal values (Wilkins, 1965)

1.40

1.20

1.00

0.80

♀

2   6   10   14   18   22   26

→ age (years)

Fig. 6. Upper/lower segment ratios of female patients.

29

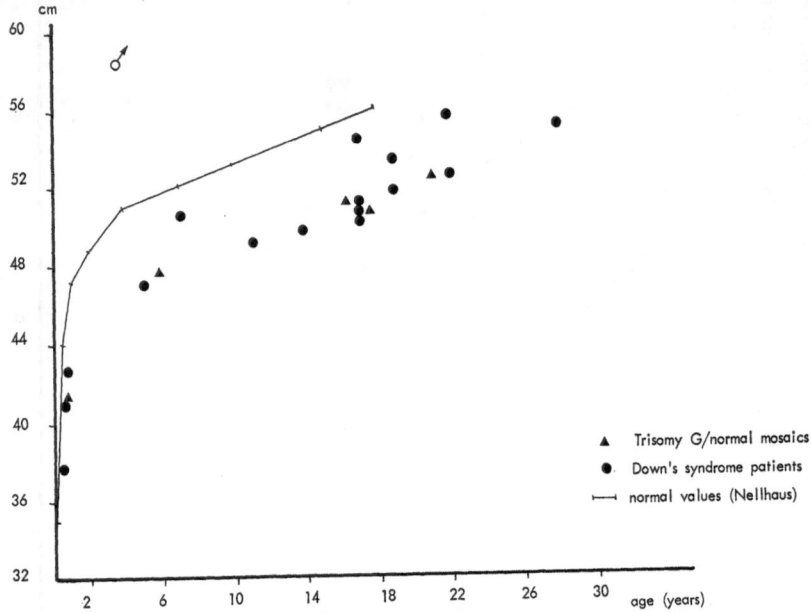

Fig. 7. Head circumferences of male patients.

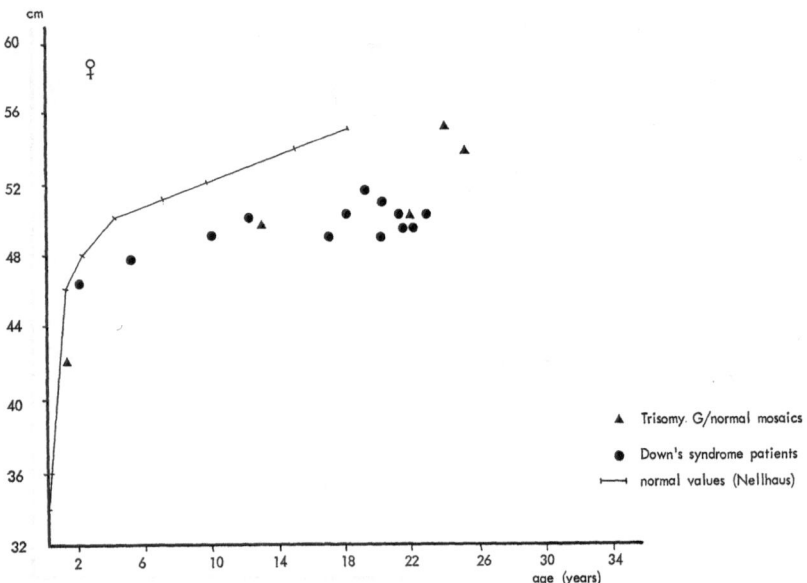

Fig. 8. Head circumferences of female patients.

30

latter also have severe reduction of total height and head circumference, and show the characteristic shortening of the lower extremities typical for Down's syndrome patients as compared with other mental defectives.

2. Characteristic non-measurable signs of Down's syndrome

a. Introduction
Several symptoms are considered characteristic of Down's syndrome by various authors. They have tried to evaluate the diagnostic significance of these signs by investigating their incidence in Down's syndrome patients. A survey of the published data is given in Table VI.

We selected several characteristics signs of Down's syndrome and investigated their incidence in the trisomy G/normal mosaic patients, 30 Down's syndrome patients, and also in 80 controls from 1 to 5 years old from a welfare clinic. The occurrence of epicanthic folds, which disappear in normal persons (Lowe 1949), was also checked in 40 adults. The number of positive signs found and their distribution according to age will be discussed.

b. Selected characteristic signs of Down's syndrome
Epicanthic folds arising from the inner part of the pars orbitalis of the eyes are frequently seen in Down's syndrome patients as well as in normal Dutch children. These folds disappear in normal persons with the growth of the face, according to Lowe (1949); they become less pronounced in older patients with Down's syndrome.

Brushfield's spots, a white speckling on the periphery of the iris, are observed in a high percentage of Down's syndrome patients. Donaldson (1961) photographed the irides of 180 mongoloids, 146 normal persons, and 12 presumably normal infants. Brushfield's spots had been present from birth in 85 per cent of the patients with Down's syndrome; they occurred in both blue and brown irides and were located more toward the midzone, being more numerous and distinct than the Brushfield's spots seen in 24 per cent of his controls.

Dysplastic ears have been described as one of the ten most important signs in newborns by Hall (1964). An overhanging helix is the most frequently seen abnormality of the ears in patients with Down's syndrome.

A furrowed tongue has been described as a pathognomonic change in

31

Down's syndrome by Engler (1949), who observed this sign in all of his patients older than 5 years. He noticed that furrowing of the tongue is not present at birth, but develops after the second year.

Abnormal teeth caused by disturbed dental development in Down's syndrome have been described by Cohen and Winer (1965). In a series

Table VI. Incidence of the characteristic non-measurable signs of Down's syndrome, as reported in the literature (in percentages)

	Øster All ages	Gustavson 1-20 yr	Levinson All ages	Engler Young patients	Donaldson	Lowe All ages	Hall Newborns	Cohen & W. 3-30 yr
Epicanthus	28	54.5	50	50—60				
Oblique palpebral fiss.	75	86.1	88					
Brushfield's spots	70	69.5	30		85	90	42	
Abnormal ears	49	28	48				23	54.2
Highly arched palate	67	69.5	74					
Furrowed tongue	59	43.6	44	100 (> 5 yr)				
Abnormal teeth	71	64.8	68	100				73
Flat occiput	74		82					
Short neck	39		50					
Short broad hands	69	74.7	74					
Four finger crease	43	60.2	48					
Curved digit V	48	52	68					
Plantar furrow			28					
Hyperextensibility	47	84.8						
Cutis marmorata	43		32					

of 168 patients, 73 per cent showed one of the following disorders: hypoplasia of teeth, white spots, semilunar notching, peg-shaped teeth, missing laterals. Barkla (1966) described a high incidence of congenital absence of the lateral incisor and second bicuspid in a series of 468 Down's syndrome patients. Absence of caries in mongolism was demonstrated by Cohen and Winer (1965), who found 53.1 per cent of Down's syndrome patients to be caries-free, and by Wash (in Benda 1969) who found no caries in 39 out of 84 cases. Benda (1969) has stated that mongoloid teeth are smaller than normal.

A short neck is difficult to define in newborns, but in older patients with Down's syndrome it clearly contributes to their characteristic appearance. Levinson et al. (1955) reported it in 50 per cent of Down's syndrome patients.

A curved fifth finger occurred in 52 per cent of Gustavson's (1964) patients with Down's syndrome, but also in 37 per cent of his patients with suspected Down's syndrome and normal karyotypes. Therefore this sign, though it seldom occurs in normal people, may also be present in other patients with mental retardation, and its diagnostic value should always be evaluated in combination with other traits of Down's syndrome.

A plantar furrow on the sole of the feet between the first and second toe was seen in 28 per cent of the patients of Levinson et al. (1955).

Cutis marmorata was found by Øster (1953) in 43 per cent of his patients and in 32 per cent of the patients of Levinson et al. (1955).

Scant body hair: Benda (1969) described absence of axillary hair, and Penrose and Smith (1966) reported scant axillary hair in patients with Down's syndrome.

c. Results of the present study
The incidence of these non-measurable signs of Down's syndrome is shown in Table VII, their occurrence in our trisomy G/normal mosaics, Down's syndrome patients and controls in Table VIII and Fig. 9. The

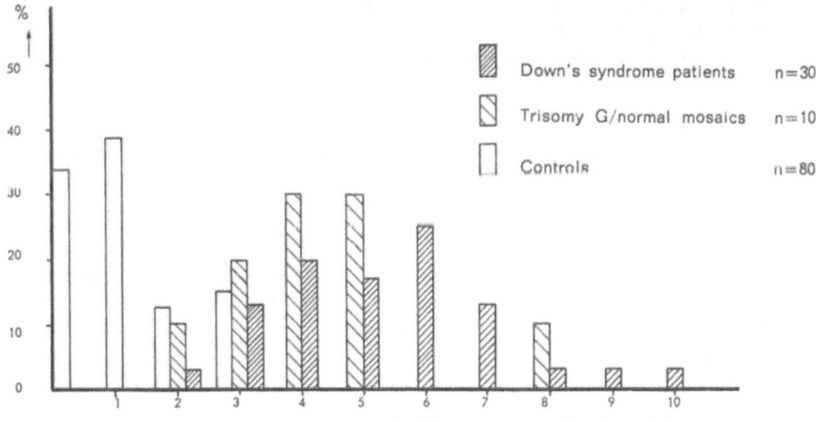

Fig. 9. Incidence of positive characteristic non-measurable signs of Down's syndrome.

mosaic patients have been numbered in all our investigations according to their percentage of trisomic leukocytes, patient 1 having the lowest percentage.

33

Epicanthic folds occurred in 38 per cent and 2.5 per cent of 80 controls from 1 tot 5 years and 40 adults, respectively, confirming the disappearance of epicanthic folds in normal persons (Lowe 1949). Epicanthic folds occurred in 53 per cent of the group of 30 Down's syndrome patients and also in 53 per cent of the patients older than 19 years, which indicates the persistence of epicanthic folds in this syndrome, contrary to the findings of Benda (1969) and Engler (1949), who found disappearance

Table VII. Incidence of the characteristic non-measurable signs
of Down's syndrome (in percentages)

Trisomy G/normal mosaics n = 10		Down's syndrome patients n = 30	Controls	
			n = 80 1—5 years	n = 40 adults
Epicanthus	60	53	38	2.5
Brushfield's spots	44	53	10	
Abnormal ears	40	47	5	
Furrowed tongue	40	83	1	
Abnormal teeth	67	75	0	
Short neck	60	40	25	
Curved digit V	50	59	9	
Plantar furrow	40	48	5	
Cutis marmorata	20	60	16	
Scant body hair	29	42	—	

Table VIII. Incidence of the characteristic non-measurable signs of Down's
syndrome in the trisomy G/normal mosaic patients

Patient No.	1	2	3	4	5	6	7	8	9	10
Sex	f	m	m	f	f	m	f	m	f	m
Age (yr)	24	19	16	21	25	21	12½	5/12	10/12	6
Epicanthus	—	+	—	+/—	—	—	+	+	+	+
Brushfield's spots	—	—	+	—	—	—	.	+	+	+
Abnormal ears	+	—	—	+	—	+	—	—	—	+
Furrowed tongue	+	—	—	—	—	+	+	—	—	+
Abnormal teeth	+	+	+	—	—	+	+	.	—	+
Short neck	—	—	+	+	+	—	—	+	+	+
Curved digit V	—	+	+	—	—	+	—	+	—	+
Plantar furrow	—	—	+	—	—	—	—	+	+	+
Cutis marmorata	—	—	—	+	—	—	—	—	+	—
Scant body hair	+	—	—	—	+	—	—	.	.	.
Number of positive signs	4	3	5	4	2	4	3	5	5	8

mean: 4.3 (2-8)

of the folds with advancing age. Of our 10 trisomy G/normal mosaics, epicanthic folds were seen in 6 patients, 3 of whom were under 10 years of age, and 3 over 10 years.

Brushfield's spots occurred in our Down's syndrome patients and controls, in 53 and 10 per cent respectively, and in 4 of 9 trisomy G/normal mosaic patients.

Abnormality of the ears was taken as the presence of a double helix or the absence of the helix, strong outward curvature of the upper part of the ears, dissimilarity between the two ears, and low implantation. Missing lobules were not considered abnormal in our series, since they are frequently seen in normal patients. Abnormal ears were seen in 47 per cent of our Down's syndrome patients, 5 per cent of the controls, and in 4 of the 10 trisomy G/normal mosaic patients.

Furrowing of the tongue was present in 83 per cent of our Down's syndrome patients, one of whom was only 14 months old. This was the highest incidence of all the investigated signs. In 80 children aged from one to 5 years we saw furrowing of the tongue only once, in a 3 year old child. The furrowing in this child, however, appeared only in part of the tongue, whereas in Down's syndrome patients the whole tongue was involved. Furrowing of the tongue, which occurred in 4 of our 8 trisomy G/normal patients older than one year, has in the meantime also developed in patient 8 at the age of 17 months.

Abnormality of the teeth was taken as deviant size, shape, and implantation and congenitally missing teeth. It can be seen from Table VII that the symptom is a very consistent one, occurring in 75 per cent of Down's syndrome patients, whereas it was not seen in any of the controls.

Six of the 9 trisomy G/normal mosaic patients who had teeth showed dental abnormalities. Several of our patients had large teeth, contrary to Benda's (1969) description.

A short neck was present in 40 per cent of the Down's syndrome patients but also in 25 per cent of the controls. Six of the 10 trisomy G/normal mosaic patients had a short neck.

A curved 5th finger was seen in 59 per cent of our Down's syndrome patients as compared with 9 per cent of the controls. Five of the 10 trisomy G/normal mosaics had a curved 5th finger.

A plantar furrow was found in 48 per cent of our Down's syndrome patients and in only 5 per cent of the controls, for which reason we have considered this sign to have diagnostic value. Four of the 10 trisomy G/normal mosaic patients had a plantar furrow.

Cutis marmorata occurred nearly four times as frequently in the Down's syndrome patients as in the controls, 60 and 16 per cent, respectively. It was seen twice in the ten trisomy G/normal mosaic patients.

Scant axillary and pubic hair was present in 42 per cent of our Down's syndrome patients and in 2 of the 7 trisomy G/normal mosaic patients who were older than 12 years.

Patient 4 with trisomy G/normal mosaicism has the symptoms of the Rubinstein-Taybi (1963) syndrome, consisting of mental retardation, antimongoloid slant of the eyes, beaked nose, and broad end-phalanxes of thumbs and great toes.

Non-measurable clinical signs of Down's syndrome occurred in 10 per cent or less of the controls with the exception of the short neck and cutis marmorata, and in 40 per cent or more of Down's syndrome patients. Epicanthic folds have little diagnostic value in children under 5 years of age.

A furrowed tongue and abnormal teeth were seen in 83 and 75 per cent, respectively, of the Down's syndrome patients and in 1 and 0 per cent of the controls, respectively, which makes combination of these symptoms in one patient important for the confirmation of the diagnosis Down's syndrome. In the trisomy G/normal mosaic patients the non-measurable signs were present in about the same percentage as that of Down's syndrome patients with the exception of a furrowed tongue and cutis marmorata, which were present in lower percentages.

The importance of the presence of the positive signs of Down's syndrome in individual patients requires discussion. The non-measurable positive signs were added up for each individual of the group of trisomy G/normal mosaics, Down's syndrome patients, and controls; the percentages of individuals in each group for positive signs are given in Fig. 9.

The controls were 1 to 5 years of age and should be seen as controls for Down's syndrome patients and trisomy G/normal mosaics of this age only. The controls have a possible maximum of 9 positive signs, since scant body hair can of course only be measured after the onset of puberty. Their incidence of positive signs can be expected to be higher than that of children under one year, since a furrowed tongue and abnormal teeth do not yet occur at this latter age. The incidence of positive signs in older children and adults will not be higher, however, because epicanthic folds are seen frequently in children up to 5 years old, whereas they only occurred in one out of 40 investigated adult patients. These results suggest that the average number of positive signs will be about the same in the control children and in normal persons at all ages.

The results show that 72.5 per cent of the controls have no or only one positive sign, and none has more than 3 positive signs (average 1.1). All the patients with Down's syndrome have 2 or more positive signs and 84 per cent more than 3. The average of this group is 5.3, the range 2-10

The trisomy G/normal mosaic patients have a range of 2-8 positive signs with an average of 4.3. The 2 patients under one year lie well within mongoloid range, each having 5 positive signs. The 6 year old mosaic patient has 8 positive characteristics of Down's syndrome and of the 7 trisomy G/normal mosaics of 12 years and older, 4 fall within mongoloid range (more than 3 signs) and 3 are in the overlapping zone (2 or 3 positive signs).

In 4 of the 10 trisomy G/normal mosaic patients (cases 1, 2, 4, and 5) the diagnosis Down's syndrome was not suspected. In 3 of these patients (cases 1, 2, 4) Down's syndrome could have been suspected because of the presence of 3 or more positive clinical signs.

It is clear from the foregoing that the number of positive clinical signs in trisomy G/normal mosaic patients varies greatly, but that the majority of these patients fall well within the mongoloid range. The intermediate place of trisomy G/normal mosaics between patients with Down's syndrome and normals is confirmed, though there is no linear relation between the percentage of trisomic cells and the number of positive signs in individual patients.

3. Dermatoglyphs

Dermatoglyphic patterns are of great value in establishing the presence of chromosomal disorders. The characteristic dermatoglyphs of Down's syndrome patients were first demonstrated by Cummins in 1936. The literature on dermatoglyphs of trisomy G/normal mosaic patients is reviewed in the literature: II, 2.

The dermatoglyphs of both our trisomy G/normal mosaic patients and their parents were analysed. The parents were investigated because Penrose (1954) described a statistically significant tendency in mothers and sibs of patients with Down's syndrome to have high positions of the palmar triradius. The dermatoglyphs of the Down's syndrome patients were also investigated and the logarithmic indices of the dermatoglyphs calculated according to Ford Walker (1958), who investigated the incidence of certain digital, plantar, and hallucal patterns in Down's syndrome patients and compared the frequencies with those of a control group of normal persons; for each pattern, the ratio of the frequencies of the patients to those of the controls was calculated and expressed logarithmically. The sum of the logarithms of these ratios in one person is the logarithmic index of his dermatoglyphs. The digital patterns of the controls of Ford Walker are similar to those of the Dutch controls (Dankmeijer 1934).

Distribution curves of log indices of Down's syndrome patients show that 71.5 per cent of these patients have an index higher than +3 and of the normal controls 68.5 per cent have an index below –3. The overlap of the two groups is remarkably small. Therefore, after the karyogram the log index is the most discriminating single sign in the diagnosis of Down's syndrome.

a. Methods

Digital, palmar, and hallucal patterns have been investigated and expressed as logarithms according to Ford Walker (1958). Dermatoglyphs can be examined with the naked eye or with a magnifying glass (x 13). Prints can be made with the printing paper and cushion of crime-detection equipment (Faurot, Inc., New York, N.Y., U.S.A.). The digital patterns can consist of a loop, which is either ulnar or radial, an arch or a whorl (Fig. 10). The patterns are made up of ridges (Fig. 11). The meeting point of three dermal ridges is called a triradius when the angle between two ridges is 120°. One triradius is present in the case of a loop, two in the case of a whorl, none in the case of an arch. A double loop is classified as a whorl. The digital patterns of Down's syndrome patients show an over-all surplus of ulnar loops. Radial loops occur in normal persons most frequently on digit II, but in Down's syndrome are more often seen on digits IV and V.

The palmar parameters used in Ford Walker's system are the site of the axial triradius and the presence of a pattern in the 3rd interdigital area IDR III (Fig. 10).

A palmar triradius is called low (t), intermediate (t'), or high (t''), according to its distance to the distal wrist crease. This distance is calcu-

38

lated as a percentage of the palm length, which reaches from the distal wrist crease to the proximal crease of the middle finger. A t lies on the proximal part of this line (occupying 0-14.9 per cent), a t' on the intermediate part (15-39.9 per cent) and the distance of a t″ to the distal wrist crease occupies 40 per cent or more of this line. A high axial triradius, t″, formed by the presence of a hypothenar pattern, is one of the most characteristic features of mongolism though by no means pathognomonic. It is also present in other trisomics and occasionally in normals.

Digital and Palmar Patterns   Hallucal Patterns

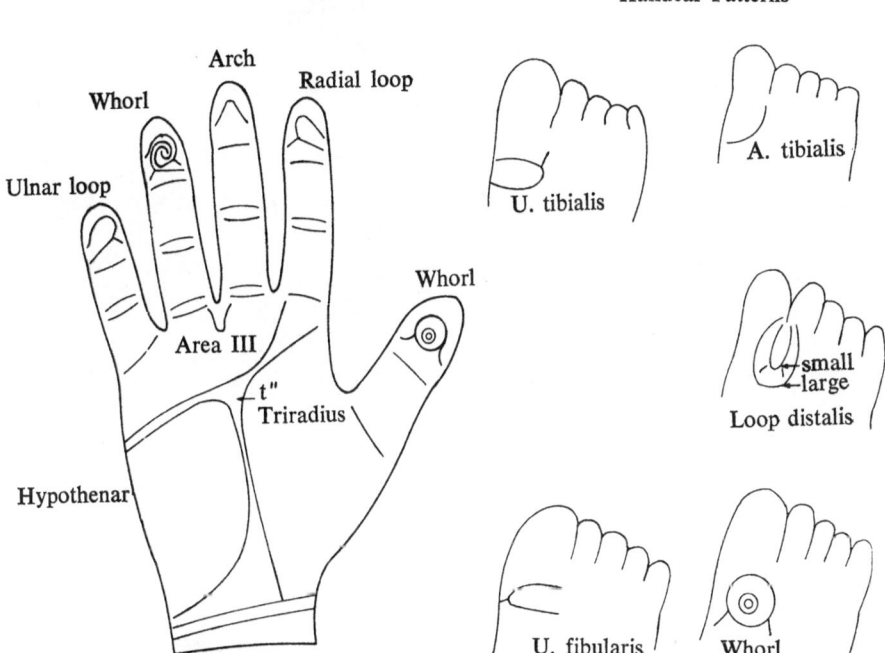

Fig. 10. Schematic representation of dermatoglyphic patterns.

A pattern in the 3rd interdigital area occurs more frequently in patients with Down's syndrome than in controls, the ratio mongol/control for left and right palms being 1.7 and 1.5, respectively (Ford Walker).

Hallucal patterns (Fig. 10) have a highly characteristic distribution in Down's syndrome patients. An arcus tibialis occurs frequently, i.e. on 46.6 per cent of left soles and 47.4 per cent of right soles, compared with an occurrence of only 0.3 per cent on both feet of controls (Ford Walker 1958). A hallucal pattern consisting of a small distal loop with 1-20 ridges from the centre of the pattern to the nearest triradius occurs

39

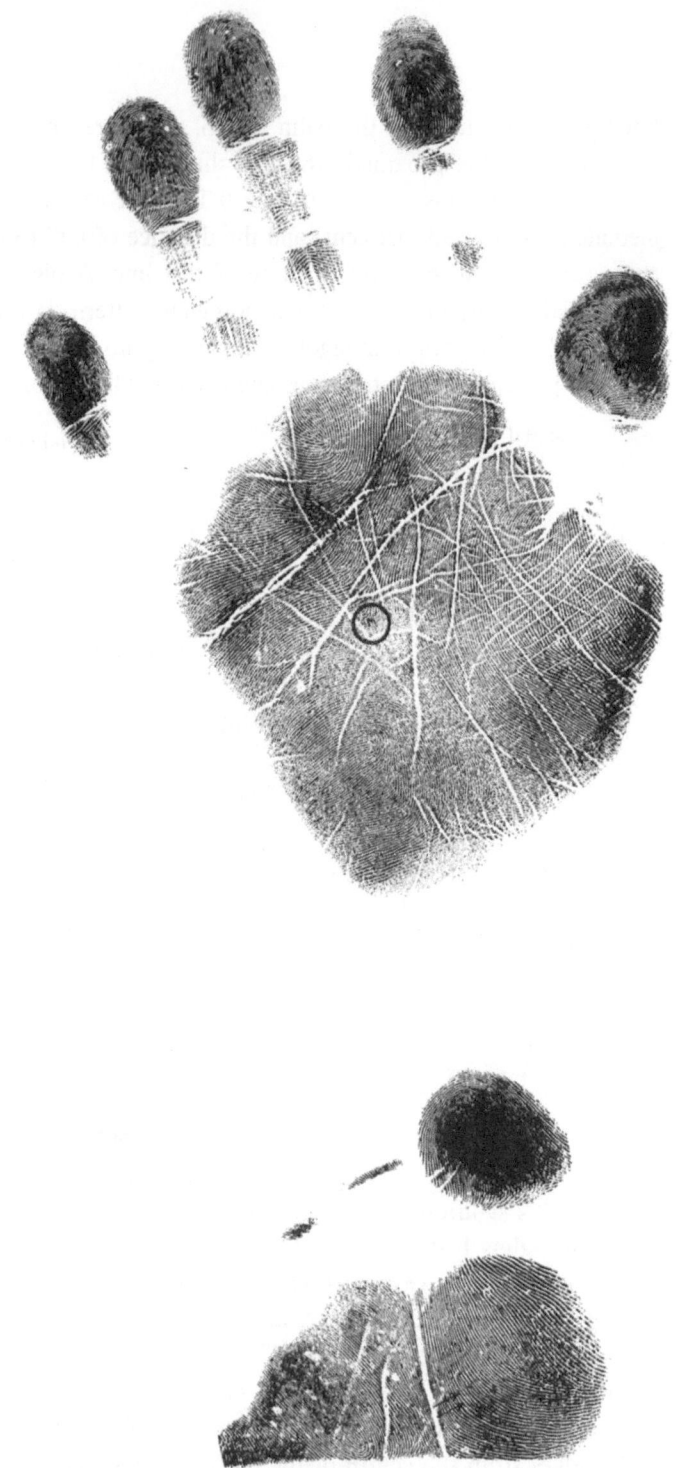

Fig. 11. Prints of left hand and foot of patient 8 (Table XI).   ○ Triradius t''
→ A. tibialis

in Down's syndrome patients in 33.8 per cent of left and 31.2 per cent of right soles, as compared with 10.0 and 13.3 per cent, respectively, of controls (Ford Walker 1958).

Logarithmic indices of our trisomy G mosaic patients, their parents, and the Down's syndrome patients, were calculated by addition of the single scores obtained for digital, palmar, and hallucal patterns of both hands and feet, according to the logarithms calculated by Ford Walker (1958). The prints of the left hand and hallucal area of patient 8 (see Table XI) are given in Fig. 11. The calculation of the log index of this patient is as follows:

Calculation of log index (Ford Walker), patient 8 (Table XI):

Left hand

Digits 5 x U	=	+0.06	+0.36	+0.07	—0.01	—0.05	=	+0.43
Trirad. t″	=		+0.92					+0.92
IDR III +	=	+0.15						+0.15

Right hand	=	+0.09	+0.42	+0.09	+0.06	—0.05	=	+0.61
Digits 5 x U	=	+0.89						+0.89
Trirad. t″	=	+0.18						+0.18
IDR III +								

Hallucal pattern

Left A.tib.	=	+2.19		+2.19
Right A. tib.	=	+2.20		+2.20

Log index = +7.57 +

b. Results

The dermatoglyphs of the trisomy G/normal mosaics, their parents, and the Down's syndrome patients, are shown in Tables IX, X, XI. A schematic drawing of the dermatoglyphs of the trisomy G/normal mosaics is given in Fig. 12. The data for controls are taken from Ford Walker (1958).

Digital patterns. The trisomy G/normal mosaics have digital patterns differing from those of the patients with Down's syndrome, since 10 ulnar loops occurred in only one mosaic patient as against 40 per cent of the Down's syndrome patients. The occurrence of ulnar loops on different digits in controls varies from 58.7 to 86.4 per cent.

Palmar patterns. A pattern in the 3rd interdigital area was found in either hand in 8 of 10 trisomy G/normal mosaic patients and in 80 per

Table IX. Dermatoglyphs of the trisomy G/normal mosaic patients

Pat. No.	Sex	Fingertips Left (5 4 3 2 1)	Fingertips Right (1 2 3 4 5)	Area III L	Area III R	Tri-radius L	Tri-radius R	Hallucal Left	Hallucal Right	Log Index Ford Walker
1	F	U U U W U	U R R U U	—	+	t	t	U.tib.	U.tib.	— 4.93
2	M	U W U W U	U A U W W	—	—	t″	t″	L.L.D.	L.L.D.	— 1.24
3	M	U U U A U	U U U U U	+	+	t″	t″	L.L.D.	A.tib.	+ 3.99
4	F	U W R A A	A A U U U	—	—	t′	t	S.L.D.	U.tib.	— 4.42
5	F	U U U A U	U W U U U	+	+	t″	t″	W	W	— 1.01
6	M	U W U U U	U U U U U	+	+	t	t	A.tib.	A.tib.	+ 4.32
7	F	U U U U U	U U U U U	+	+	t″	t″	A.tib.	A.tib.	+ 7.57
8	M	U W U U A	A U W W U	—	+	t″	t″	S.L.D.	S.L.D.	+ 2.89
9	F	U U U U U	U U U A U/A	+	+	t″	t	A.tib.	A.tib	+ 6.24
10	M	U W U U U	U U U U U	+	—	t″	t″	A.tib.	A.tib.	+ 6.83

U = Ulnar loop   t = low   U.tib. = Tibial loop
R = Radial loop   t′ = intermediate   A.tib. = Arch tibial
A = Arch   t″ = high   L.L.D. = Large loop distalis
W = Whorl   S.L.D. = Small loop distalis

Table X. Dermatoglyphs of the parents of the trisomy G/normal mosaic patients

Pat. No.	Sex	Fingertips Left (5 4 3 2 1)	Fingertips Right (1 2 3 4 5)	Area III L	Area III R	Tri-radius L	Tri-radius R	Hallucal Left	Hallucal Right	Log Index Ford Walker
1	P	U U U W U	U R R U U	+	+	t	t	L.L.D.	L.L.D.	— 3.97
	M	U U A A U	U A A U U	—	+	t	t	L.L.D.	L.L.D.	— 3.45
2	P	U U U W U	U W U W U	+	+	t″	t″	L.L.D.	L.L.D.	+ 0.31
	M	U U A R U	U U U U U	—	—	t	t′	L.L.D.	L.L.D.	— 4.23
3	P	U U U A U	U A U U U	—	+	t	t	L.L.D.	S.L.D.	— 2.83
	M	U A A A A	A A A U U	+	+	t	t	S.L.D.	S.L.D.	— 1.26
4	P	U U U R U	U R W W U	—	—	t	t	W	L.L.D.	— 6.03
	M	U U A R U	W U U U U	—	—	t	t	W	L.L.D.	— 5.49
5	P	U U U R U	U W U W U	+	—	t	t	W	W	— 5.56
	M	U U U A U	U U A U U	—	—	t	t	W	W	— 3.18
6	P	U U U U U	U W U U U	+	+	t	t	L.L.D.	L.L.D.	— 1.87
	M	U W U U U	U W U W W	—	+	t′	t	L.L.D.	L.L.D.	— 2.28
7	P	—	—	—	—	—	—	—	—	—
	M	U U U U W	W A U U U	—	—	t	t	—	—	—
8	P	U U A R U	U R A U U	—	+	t	t	S.L.D.	S.L.D.	— 1.98
	M	U W W W A	U W W W U	—	+	t	t	A.fib.	L.L.D.	— 4.45
9	P	U U U U W	W W U U U	—	+	t	t	W	S.L.D.	— 2.75
	M	U W U U U	U W U W U	+	+	t	t	L.L.D.	L.L.D.	— 2.14
10	P	—	—	—	—	—	—	—	—	—
	M	W W U R W	W R U W W	—	—	t	t	L.L.D.	L.L.D.	— 5.21

Table XI. Dermatoglyphs of the 30 Down's syndrome patients

Pat. No.	Sex	Hands Fingertips Left 5 4 3 2 1	Right 1 2 3 4 5	Palms Area III L R	Tri- radius L R	Feet Hallucal pattern Left	Right	Log Index Ford Walker
1	M	U U U U U	U U U U U	+ +	t t	S.L.D.	S.L.D.	+ 1.86
2	M	U U U W U	U U U U U	— +	t″ t″	A.tib.	A.tib.	+ 6.47
3	F	U W U U U	U U U W W	+ +	t″ t″	S.L.D.	S.L.D.	+ 3.96
4	F	U W U U U	U U U U U	+ +	t t″	S.L.D.	S.L.D.	+ 2.41
5	M	— — — A —	— A A A —	+ +	t″ —	A.tib.	A.tib.	>+ 2.78
6	M	U U U U U	U U U U U	+ —	t″ t″	S.L.D.	S.L.D.	+ 3.37
7	F	U U U U U	U U U U U	+ +	t t	S.L.D.	S.L.D.	+ 0.86
8	M	U U U U U	U U U U U	+ +	t″ t″	A.tib.	A.tib.	+ 7.57
9	F	U U U U U	U U U U U	+ +	t t	A.tib.	A.tib.	+ 4.26
10	M	U W U U U	W U U U U	— —	t″ t″	A.tib.	A.tib.	+ 6.29
11	F	W W U U W	W U U W W	— —	t″ t″	A.tib.	S.L.D.	+ 5.19
12	F	U U U U U	U U U U U	+ +	t t	L.L.D.	L.L.D.	— 1.07
13	F	U U U U U	U U U U U	+ +	t′ t′	A.tib.	S.L.D.	+ 2.68
14	M	U U U U U	U U U A U	— —	t t″	L.L.D.	L.L.D.	— 0.19
15	M	U W U U U	U U U U U	— +	t′ t′	S.L.D.	S.L.D.	+ 0.52
16	F	U W U U W	U U U U U	+ +	t″ t	L.L.D.	L.L.D.	+ 0.31
17	M	U U U U W	U U U U U	+ +	t t	S.L.D.	S.L.D.	+ 0.66
18	F	U U U U U	A U U R U	— —	t″ t″	A.tib.	A.tib.	+ 8.43
19	F	W W U U U	U U U W W	+ +	t″ t″	A.tib.	A.tib.	+ 7.69
20	M	U U U U U	W U U U U	+ +	t t	A.tib.	S.L.D.	+ 2.26
21	F	U W U U W	U U U W U	— —	t″ t″	S.L.D.	S.L.D.	+ 2.60
22	M	R U U U W	U U U U W	+ +	t t	S.L.D.	A.tib.	+ 3.84
23	M	U U U U U	U U U U U	+ +	t″ t″	A.tib.	A.tib.	+ 7.57
24	M	U U U U U	U U U U U	+ +	t″ t″	A.tib.	A.tib.	+ 7.57
25	F	U U U A U	U U U U U	+ +	t″ t″	A.tib.	A.tib.	+ 6.70
26	M	U U U U U	U U U U U	+ +	t″ t″	S.L.D.	S.L.D.	+ 4.07
27	M	U W U U U	U U U U U	— —	— t″	A.tib.	A.tib.	>+ 4.91
28	F	U U U U U	U W U U U	+ +	t″ t″	L.L.D.	L.L.D.	+ 1.34
29	M	U U U U U	U U U U U	+ +	t′ t	A.tib.	A.tib.	+ 4.36
30	M	U U U U U	U U U U U	+ +	t t	A.tib.	A.tib.	+ 4.36

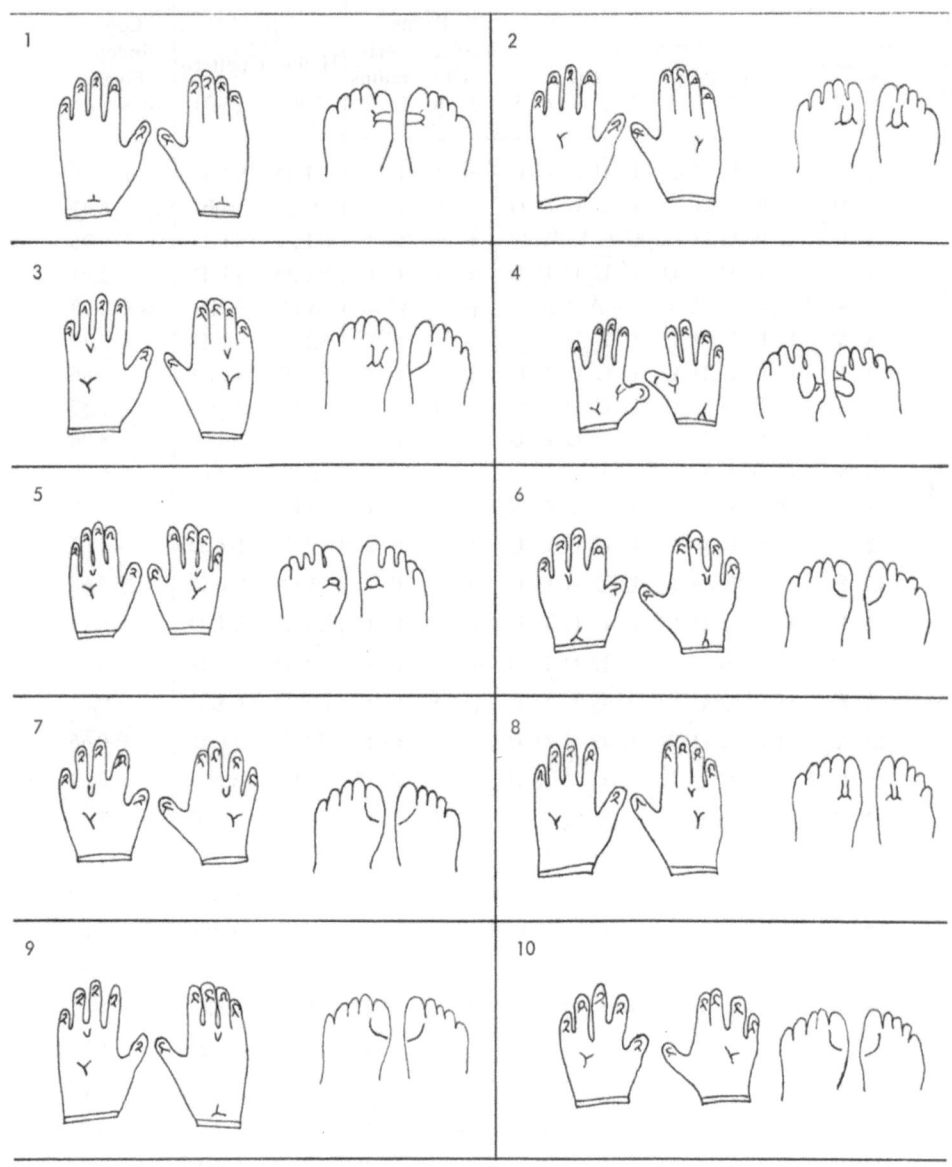

Fig. 12. Dermatoglyphs of the trisomy G/normal mosaic patients.

cent of the patients with Down's syndrome. For the controls these values are for left and right hands 31 and 55 per cent, respectively. A high palmar triradius was present in left and right hands of 7 and 6 mosaic patients and 55 and 59 per cent of Down's syndrome patients, respectively. For the controls, these values are 10 and 13 per cent, respectively.

Hallucal patterns. The characteristic arcus tibialis of Down's syndrome was found in 9 out of 20 feet of the mosaic patients and in 50 per cent of the feet of the Down's syndrome patients; the incidence in controls is 0.3 per cent.

Logarithmic indices. A histogram of the logarithmic indices of the dermatoglyphs of the trisomy G/normal mosaics and Down's syndrome patients is given in Fig. 13. The mosaic patients nos. 6 to 10, with more than 30 per cent trisomic leukocytes, all have an index within mongoloid range. The same applies to patient 3, with only 13 per cent trisomic leukocytes but 92 per cent trisomic fibroblasts. Patients 1 and 4 have a logarithmic index well within normal range, and patients 2 and 5 can be considered to have indices in the intermediate range, in both cases due to the high position of the palmar triradius. Patient 4 with the Rubinstein-Taybi syndrome shows a surplus of digital arches, as described by Giroux and Miller (1967), and a pattern in the thenar/first interdigital area demonstrated by Robinson et al. (1966) to be a consistent feature of this syndrome.

A negative log index was found for only 2 patients with Down's syndrome, a frequency in accordance with the frequency of negative indices found by Ford Walker (1958). Patients 11 and 30 with Down's syndrome are the D-G translocation patients. Their dermatoglyphs do not differ essentially from those of the trisomy G patients, which is in agreement with Rosner and Ong's (1967) results.

Table X shows the distribution of the dermatoglyphs of the parents of the trisomy G/normal mosaics. A deviation of the palmar triradius of mothers of mongols toward that of mongols, possibly caused by mosaicism in these mothers, was described by Penrose in 1954. Normal positions of the palmar triradius were found in the mothers of our patients, and only one of 8 fathers had a high triradius in both hands, which is a normal incidence. All these parents except one had log indices well within the normal range.

It may therefore be said that a percentage above 30 of trisomic cells in either blood or skin is reflected in positive log indices in these patients.

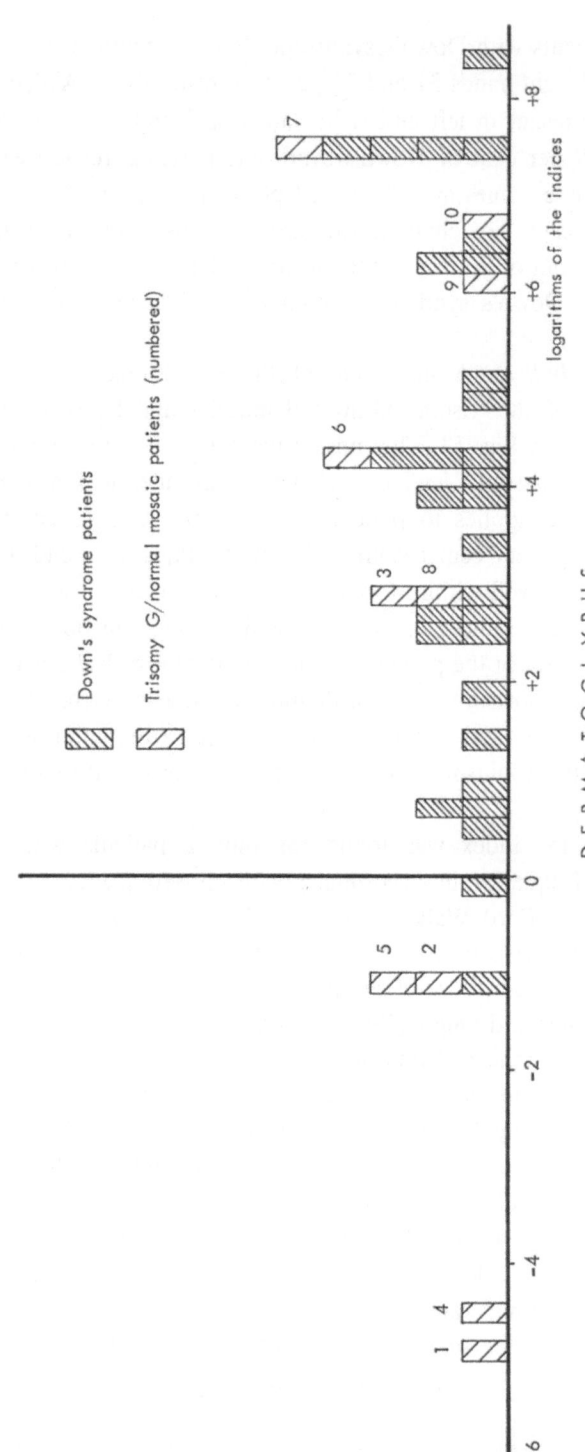

Fig. 13. Histograms of indices based on digital, palmar, and plantar patterns.

The positive values of the log index of 6 mosaic patients and the intermediate values of 2 mosaic patients are caused by the large number of high palmar triradii and the frequent presence of the arcus tibialis.

4. Radiology of pelvic bones

Caffey and Ross (1956) were the first to report changes of the pelvic bones in infants with Down's syndrome. These changes consist of small acetabular angles and widening of the iliac wing and bodies, resulting in small iliac angles. The iliac index – the sum of both acetabular and iliac angles divided by 2 – was consequently much smaller in these infants than in normals. They found X-ray findings to be diagnostic in 4 out of 5 mongols. Their examinations and those of various other authors (Table XII) concerned newborns or children below the age of one year. Additional data on children aged 4 to 18 years were collected by van Gelderen (1967). The acetabular angles of his older controls are much smaller than those found by other authors for controls under one year, and the mean acetabular angle larger for girls than for boys. The iliac angle shows an increase with age for both girls and boys. The results of the various authors are summarized in Table XII.

Investigations of the pelvic bones of trisomy G/normal mosaic patients have not yet been reported. It seemed of interest to take pelvic measurements in our mosaic patients and compare the results with those of patients with Down's syndrome and of controls, matched for sex and age.

a. Methods

Acetabular angles were measured according to Hilgenreiner (1925). The Y-Y line is drawn through the 2 Y cartilages. A second line connecting the two caudal points of the ilium is drawn. The angle made by these two lines is the acetabular angle (Fig. 14). Acetabular angles can only be measured before the calcification of the Y-Y cartilage has been completed.

Iliac angles were measured according to Ross as described by Caffey and Ross (1958). The Y-Y line is drawn. Then lines are drawn connecting the lateral end of the lower edge of the ilium with the outermost lateral point on the wing. The angle made by these two lines is the iliac angle. When it is difficult to draw the Y-Y line, the lines

Table XII. Acetabular and iliac angles, as reported in the literature

Author	Age	Down's syndrome patients						Normals					
		Acetab. angle		Iliac angle		Iliac index		Acetab. angle		Iliac angle		Iliac index	
		mean	S.D.	mean	S.D.	mean	S.D.	mean	S.D.	mean	S.D.	mean	S.D.
Caffey et al. 1958	0- 3 months	16	4.5	44	6.5	60	9.9	28	4.7	55	5.5	81	8.0
idem	3-12 months	11	4.2	41	7.0	50	9.6	22	4.2	58	7.0	79	9.0
Hall 1964	newborns	—	—	—	—	65	—	—	—	—	—	88	—
idem	12 months	—	—	—	—	55	—	—	—	—	—	—	—
Nicolis et al. 1963	0-12 months	12.19	4.63	41.47	4.83	—	—	21.20	5.07	55.44	5.46	—	—
Armendares et al. 1967	newborns	—	—	—	—	64.2	8.7	—	—	—	—	83.5	5.76
Astley 1963	0-12 months	—	—	—	—	56.5	—	—	—	—	—	82	—
van Gelderen 1967	4-10 yr	—	—	—	—	—	—	M 10.1 / F 14.2	2.8 / 2.4	57.5	6.6	—	—
idem	11-18 yr	10.42	—	49.3	—	—	—	—	—	61.3	7.4	—	—
Kaufmann 1961	0-12 months	14	—	42	—	50	—	—	—	—	—	—	—
Schultze-Jena 1959	0-12 months	11	—	39	—	51	—	—	—	—	—	—	—
Over-all average	first year	13	—	42	—	56	—	24	—	56	—	83	—

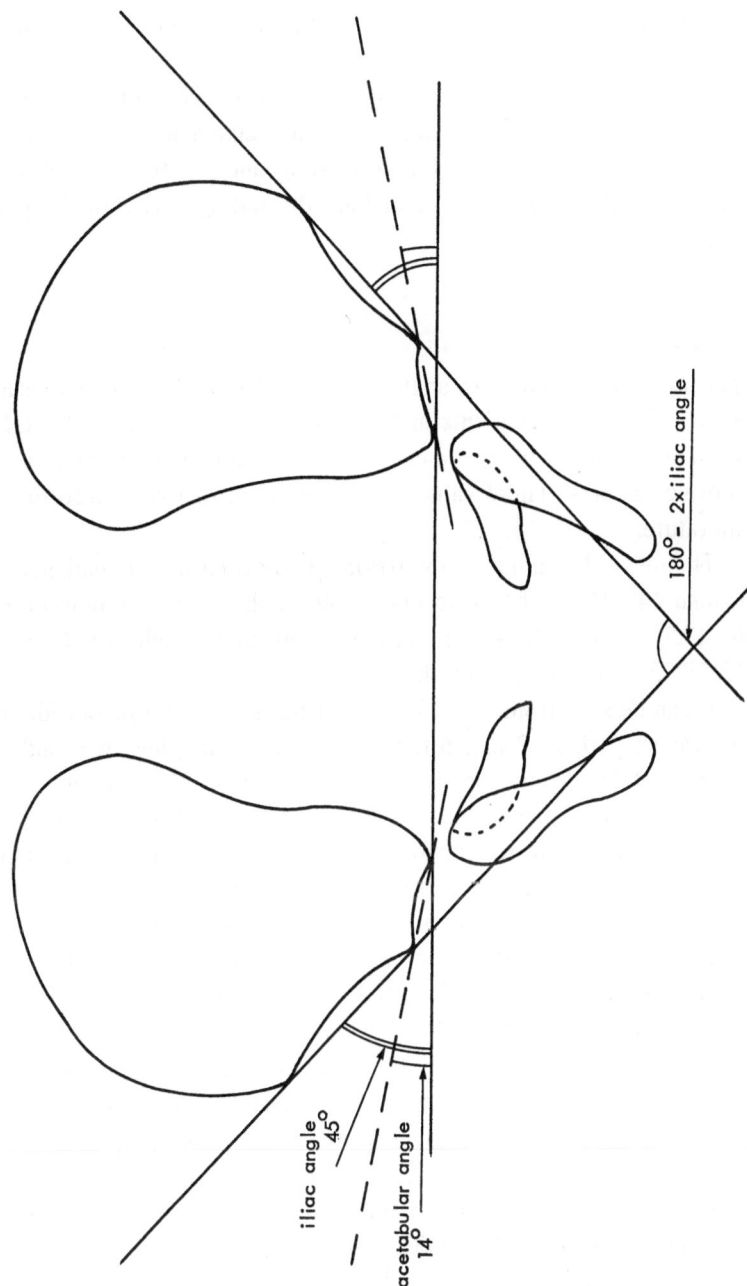

iliac angle 45°

acetabular angle 14°

180° − 2×iliac angle

Fig. 14. Schematic representation of acetabular and iliac angles.

along the iliac wing can be prolonged and the angle made by these two lines will be 180° minus two iliac angles.

Taybi and Kane (1968), who published a review of the literature on small acetabular and iliac angles and associated diseases, demonstrated a remarkable influence of forward tilting and rotation on pelvic angles, and stress the importance of making the radiograms with the patient in a neutral position.

b. Results

The acetabular and iliac angles of our trisomy G/normal mosaics, the Down's syndrome patients and controls are shown in Table XIII. Two acetabular angles of 14° and 16° were found in the trisomy G/normal mosaic patients. These values are within control range, according to the literature.

The mean iliac angle of the trisomy G/normal mosaic patients is 55.1° (range 44°-71°). The mean iliac angle of the Down's syndrome patients is 47.7° (range 40°-57°), and a mean iliac angle of 64.3° (range 53.5°-72°) was found for the controls.

It can be seen from Fig. 15 that patients 1, 2, and 5 lie outside mongoloid range, patients 3 and 6 fall within the overlapping area, and patients 4, 8, and 10 lie outside the range of the controls. Patient 4, with the Rubinstein-Taybi syndrome, has iliac angles below control range, as has also been demonstrated for four such patients by Taybi and Kane (1968). The numbering of patients is the same as in other chapters, according to their percentage of trisomic leukocytes, patient 1 having the lowest percentage. It is evident that the higher the degree of mosaicism, the lower the iliac angle, which supports the diagnostic value of iliac angles.

Iliac indices could only be calculated for two of our patients. Astley (1963) has stated that an iliac index below 60 for children under one year means that the patient very probably has Down's syndrome; an iliac index between 60 and 68 gives a 90 per cent chance of mongolism; an iliac index between 68 and 78 indicates a 6 per cent chance of mongolism, and iliac indices higher than 78 very probably mean normality.

Patient 8, with an iliac index of 65, therefore has, according to Astley, a 90 per cent chanche of mongolism, which is in good agreement with the fact that his iliac angle lies outside the control range.

It has been demonstrated that iliac angles, which can always be

Table XIII. Acetabular and iliac angles

Pat. No.	Trisomy G/normal mosaics					Down's syndrome patients					Controls				
	Sex	Age	Ace-tab. A	Iliac A	Iliac Index	Sex	Age	Ace-tab. A	Iliac A	Iliac Index	Sex	Age	Ace-tab. A	Iliac A	Iliac Index
1	F	25	—	71	—	F	23	—	43	—	F	23	—	57	—
2	M	14	14	67	81	M	14	8	40	48	M	14	18.5	53.5	72
3	M	16	—	52.5	—	M	15	—	47	—	M	16	—	67	—
4	F	20	—	44	—	F	22	—	52	—	F	20	—	60.5	—
5	F	25	—	58.5	—	F	23	—	45	—	F	24	—	72	—
6	M	20	—	55	—	M	20	—	57	—	M	20	10	70	80
7	No measurements					—	—	—	—	—	—	—	—	—	—
8	M	6/12	16	49	65	M	5/12	12	44	56	M	5	9	67	76
9	No measurements					—	—	—	—	—	—	—	—	—	—
10	M	6	—	44	—	M	6	10	53.5	63.5	M	6	16.5	67.5	84
Mean angle				55.1					47.7					64.3	
Range				44—71					40—57					53.5—72	

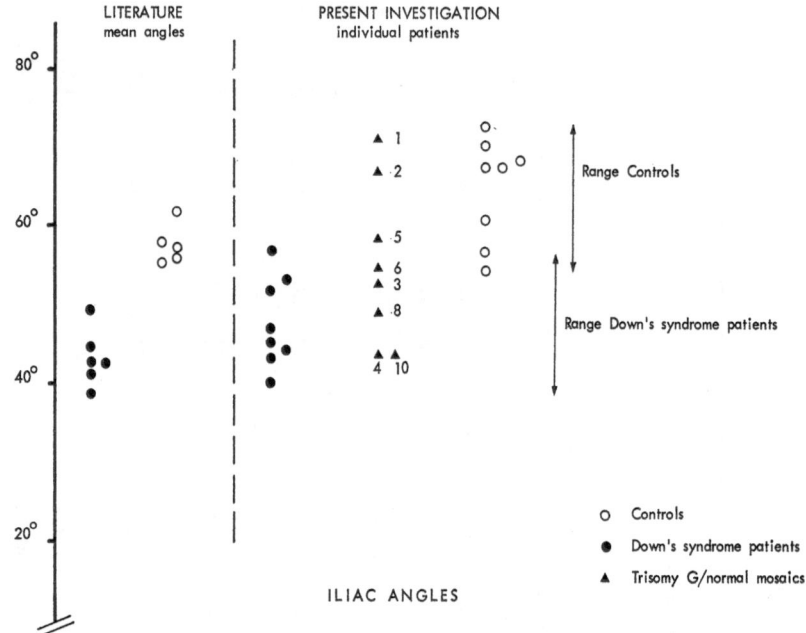

Fig. 15. Values of mean iliac angles reported in the literature and obtained for patients and controls in the present investigation.

measured, are helpful in establishing the diagnosis of partial or complete Down's syndrome.

5. Results of the clinical examinations

The data of the clinical investigations of the trisomy G/normal mosaic patients are summarized in Table XIV. Patients 1, 2, 4, and 5 with low percentages of trisomic cells and negative log indices, have some clinical signs of Down's syndrome, thus indicating the possibility of mosaicism. Patient 4 is the first patient described with both the Rubinstein-Taybi syndrome and an extra G chromosome. Patients 3, 6, 7, 8, 9, and 10, with more or less obvious signs of Down's syndrome, can only be differentiated from regular trisomy G patients by their karyogram.

Because of the selection of the cases, a possible correlation between degree of mental deficiency and mosaicism could not be detected. In this sample there certainly was no relationship whatsoever.

A direct relationship between the various investigated signs and the

Table XIV. Clinical results obtained in the trisomy G/normal mosaic patients

Patient No.	1	2	3	4	5	6	7	8	9	10
Sex - Age	F 24	M 19	M 16	F 21	F 25	M 21	F 12½	M 5/12	F 10/12	M 6
% Trisomy: Leukocytes	7	9	13	13	15	30	60	74	80	84
Fibroblasts	22	15	92	3	11	51	—	78	—	85
Body measurements within Down's syndrome range	+	+	+	+	+	+	+	+	+	+
Non-measurable characteristic signs of Down's syndrome	4	3	5	4	2	4	3	5	5	8
Dermatoglyphs log index	− 4.93	− 1.24	+ 3.99	− 4.42	− 1.01	+ 4.32	+ 7.57	+ 2.89	+ 6.24	+ 6.83
Iliac angle range:	71	67	52.5	44	58.5	55	—	49	.	44
Down's syndrome				+				+		+
Controls	+	+			+					
Overlap			+			+				

53

extra G chromosome is not apparent. However, patients with more than 30 per cent trisomic cells in either blood or skin have dermatoglyphs with positive logarithmic indices, and iliac angles tend to be lower in patients with higher percentages of trisomic cells.

IV. MATERNAL AGES AND BIRTH WEIGHTS

1. Maternal ages

It has been known since the first publication on the syndrome by Down (1866) that these patients are often born to older mothers. Richards (1969) had investigated the maternal ages of trisomy G/normal mosaic patients; the mean maternal age in his material was 31.9 for mosaic patients and 33.3 for Down's syndrome patients, and the estimated control mean 27.6. The lower mean maternal age of the mosaic patients could in his opinion be caused by the origination of a percentage of these patients from normal zygotes. The Down's syndrome patients were thought to have originated from trisomic zygotes caused by high maternal age. The mean maternal age at birth of our trisomy G/normal mosaic patients is 33.1 years (Table XV), that of the Down's syndrome patients, without the translocation patients, 33.8 years. The mean maternal age at birth for normal children born in the same years as our patients is about 29 years, 55 per cent of their mothers being younger than 30 years. Of the mothers of our trisomy G/normal mosaic patients 30 per cent were youger than 30 years, as was the case for 33 per cent of the mothers of the Down's syndrome patients.

Mean maternal ages at birth of the patients of both our groups therefore do not differ significantly (P⩾ 0.10; Wilcoxon's test) and the mean maternal age at birth of our trisomy G/normal mosaic patients, (33.1 years) is the same as that found by Richards (1969) for Down's syndrome patients.

In other words, if the higher maternal age is correlated with the occurrence of trisomy in Down's syndrome, this possibly also holds for the trisomy G/normal mosaics. The individuals in the latter group therefore presumably originated from a trisomic zygote with subsequent loss of a G chromosome in some cells.

Table XV. Trisomy G/normal mosaic patients

Birth weights (g)			Maternal ages (yr) *
1	F	1500	32
2	M	3000	43
3	M	2700	32
4	F	3250	27
5	F	3750	37
6	M	3500	32
7	F	3200	36
8	M	2800	25
9	F	3380	41
10	M	2850	28
Mean		2993	Mean 33.1

* Mean maternal age of the patients with Down's syndrome: 33.8.

2. Birth weights

The average birth weight of normal boys and girls in The Netherlands, as determined by van Gelderen in 1954, is 3530 and 3369 g, respectively. A greater variability of the birth weight in children with Down's syndrome than in normal infants was demonstrated by Levinson et al. (1955). A lower birth weight in children with Down's syndrome in The Netherlands has been described by Swaak (1967), who found a mean birth weight for Dutch boys and girls with Down's syndrome of 3141 and 3035 g, respectively. The percentage of boys and girls with birth weights of 2500 g and less was 14 and 20 per cent, respectively. In normal Dutch infants this holds for 3.5 and 5 per cent, respectively (van Gelderen 1954).

The mean birth weight of our trisomy G/normal mosaic patients was 2993 g (Table XV). When patient 1, the only premature infant, with a birth weight of 1500 g, is excluded, the mean birth weight of the mosaic patients is 3159 g, and approximates that found for Down's syndrome patients by Swaak (1967).

Since the duration of pregnancy is not well known for our cases, it cannot be determined whether the lower birth weight is caused by intra-uterine growth retardation or by a shortened gestation period.

V. LEUKOCYTES

Leukocyte changes have been described in Down's syndrome patients and trisomy G/normal mosaics (van Gelderen et al. 1967). We investigated the alkaline phosphatase levels of the neutrophil granulocytes, and morphologic studies included lobe counts of neutrophil granulocytes and fine-structural studies of their granules. The studies were carried out in the trisomy G/normal mosaic patients, and the results compared with those of Down's syndrome patients and controls, matched for sex and age.

1. Alkaline phosphatase activity - Histochemical and biochemical investigations

After the Philadelphia chromosome, which is a deleted G chromosome, had been detected in patients with chronic leukaemia (Nowell and Hungerford 1960, Baikie et al. 1960) who were known to have lowered leukocyte alkaline phosphatase levels (Valentine and Beck 1951, Valentine et al. 1957), investigations of these enzyme levels were carried out in Down's syndrome patients with the extra G chromosome. Since the average leukocyte alkaline phosphatase levels were found to be raised in these patients (Trubowitz et al. 1962) and even a ratio of 3 : 2 as compared with normals was found, a direct relationship between genetic material on the extra chromosome and leukocytic alkaline phosphatase was assumed.

In these investigations, however, there was a large overlap between normals and Down's syndrome patients, and the mean elevation of leukocyte alkaline phosphatase levels frequently showed a ratio deviating from 3 : 2. Also, many other enzyme activities proved to be increased in leukocytes of patients with Down's syndrome. If the increased alkaline phosphatase activities were directly related to the presence of the extra G chromosome in Down's syndrome patients, trisomy G/normal mosaics could be expected to show a mixed population of leukocytes with both normal and elevated indices, resulting in an average value of leukocyte alkaline phosphatase lying between that of controls and that of Down's syndrome patients.

However, previous investigations (van Gelderen et al. 1967) showed that the mean Kaplow (1955, 1968) index for a group of nine trisomy G/normal mosaic patients was even higher than that of the Down's syndrome patients. Since the Kaplow method is a semi-quantitative one based on visual estimation of the degree of positivity, it could be argued that the mean score is not a good measure of the actual mean enzyme content of the leukocytes. It is also possible that only certain iso-enzymes are demonstrated by the Kaplow method.

To investigate this, the mean alkaline phosphatase activity was determined biochemically in leukocyte suspensions prepared from the venous blood of trisomy G/normal mosaics, controls, and Down's syndrome patients, and the Kaplow indices of these persons were also determined. Kaplow index determinations were carried out at intervals of several weeks, to see how much the alkaline phosphatase content varied in time within the individual. Lobe counts of neutrophil granulocyte nuclei were made at the same time.

a. Material and methods
Human leukocyte suspensions. Human leukocytes were isolated from blood obtained by venous puncture. Fifty volumes of blood ($\pm$ 20 ml) were transferred to a glass cylinder containing 10 volumes of cooled (4° C) polyvinyl pyrrolidon solution (Isoplasma, Vifor, Geneva, Switzerland) to which one volume of heparin (400 U/ml) was added. The tube was closed and inverted carefully two or three times, after which sedimentation under gravity was allowed to proceed in the re-opened tube until a distinct separation of the red cell layer had occurred. Plasma was then pipetted off and centrifuged for 10 minutes at 200 x g, after which the supernatant fluid, which contained most of the lymphocytes and thrombocytes, was discarded. Residual plasma was removed by mixing the leukocyte pellet with 25 volumes of 0.15 M NaCl containing 0.04 M disodium ethylene diamine tetra-acetate (EDTA) adjusted to pH 7.0 with 0.1 N NaOH. The suspension was centrifuged for 10 minutes at 200 x g, after which the supernatant fluid was discarded. If necessary, selective lysis of remaining erythrocytes was performed by adding 5 volumes of cold distilled water to the pellet and stirring the cells for 10 to 20 seconds. Immediately thereafter, the suspension was readjusted to isotonicity by the addition of 5 volumes of 0.3 M NaCl. After another centrifugation for 5 minutes at 200 x g, the haemoglobin-containing supernatant fluid was discarded by aspiration. The washing procedure was repeated twice with 25 volumes of 0.15 M NaCl (pH 7.0), and the remaining leukocytes were suspended in 0.15 M NaCl in a concentration of about 2 x 10⁷ per ml. All glassware

57

was siliconized; centrifugations were carried out at 4° C.

Cell counting was performed by making a total of 8 counts in volumes of 0.1 μl each, in two Bürker counting chambers. The standard error of the mean was between 1 and 4 per cent. Differential counting of the percentage of neutrophil granulocytes was performed in cell smears made from the same cell suspension, which can also be used for the Kaplow staining.

After counting, the cell suspension was sonicated with an S-75 Branson ultrasonic disintegrator (Danbury, Conn., U.S.A.) for 5 minutes (200 Volt, stand-by 2, nominal frequency 10,000 cycles/sec), during which the suspension was kept in a crushed-ice bath at approximately 0° C and sonification was stopped for half a minute after each minute to cool the

metal tip.

Alkaline phosphatase scoring of stained cell smears. The leukocyte alkaline phosphatase index was determined by the Kaplow (1955) method in blood smears and in the smears made from the cell suspensions. The cells were counterstained for 5 minutes with Mayer's haematoxylin stain. To exclude an elevation of alkaline phosphatase caused by the infections frequently occurring in trisomy G mosaic and Down's syndrome patients, no scoring was done when the total leukocyte count exceeded 12,000 or the sedimentation rate was higher than 16.

Biochemical assay of alkaline phosphatase activity. A modification of the method of Garen and Levinthal (1960) was used: disodium P-nitro-phenyl phosphate (British Drug House, Poole, England) was dissolved in 0.001 N HCl in a final concentration of 0.038 M. A buffer was prepared from 50 volumes 0.2 M 2-amino-2-methyl-1,3-propanediol (British Drug House, Poole, England) to which added 5.7 volumes 0.2 N HCl, 0.4 volumes $MgCl_2$ (1.016 g $MgCl_2$. 6 H_2O in 10 ml distilled water) and 100 volumes distilled water. After adjustment of the pH to 9.6 (with 0.2 N HCl) the volume was brought to 200. Immediately prior to the micro-assay, 100μl of substrate solution was mixed with 100μl of buffer in appropriate polyethylene-stoppered vessels (Eppendorf, Gerätebau, Hamburg). The mixture was prewarmed in a waterbath at 37° C for 5 minutes, after which 20 μl of sonicate was added and mixed thoroughly.

After incubation periods of 20, 40, and 60 minutes, the enzymatic reaction was stopped by the addition of 200μl 0.16 M NaOH. It was found necessary to centrifuge the mixture prior to the absorbance readings at 405 nm, to remove the precipitated protein. For every determination, 4 aliquots of each of three dilutions of sonicate were assayed during three incubation periods. The slope of the relation between enzymatic activity and incubation time was calculated for each sonicate dilution by the method of the least squares (Youden 1951). By plotting these values against the leukocyte concentrations, the enzymatic activity per minute per neutrophil could be calculated.

b. Results

Mean Kaplow scores and the variations in individuals. The results of the leukocyte alkaline phosphatase scores, which were as far as possible determined at three different times in each Down's syndrome patient, trisomy G mosaic, and control, are shown in Table XVI. It is apparent that the Kaplow index for one individual varies with time and that the mean scores of the individuals within each group also vary. The ranges of the three groups show some overlapping, but the difference between the mean scores of the Down's syndrome patients and trisomy G mosaics on the one hand and those of the controls on the other remains significant ($P < 0.01$).

The mean scores are higher for trisomy G mosaics than for Down's syndrome patients, but this difference is not significant ($P > 0.50$). The P-values have been calculated by means of an analysis of variance. There is no regression with age, contrary to the findings of Alter and Lee (1968).

Kaplow scores versus biochemical alkaline phosphatase assay. In Fig. 16 the biochemically determined alkaline phosphatase activities in leuko-cytes of three trisomy G/normal mosaics, cases 2, 5, and 6, four Down's syndrome patients, and five controls are compared with Kaplow scores of the same blood samples. The correlation coëfficient between the two methods is 0.85. This very high positive correlation demonstrates the usefulness of the Kaplow method.

The three trisomy G/normal mosaics and three Down's syndrome patients show high alkaline phosphatase activities, whereas one Down's syndrome patient and all the controls have Kaplow scores within control range.

Table XVI. Kaplow scores

Pat. No.	Trisomy G/normal mosaics							Down's syndrome patients						Controls					
	Sex	Age (yr)	Scores				Mean score	Sex	Age (yr)	Scores			Mean score	Sex	Age (yr)	Scores			Mean score
1	F	25	91	96	103		97	F	18	180	292	176	216	F	21	77	110	39	75
2	M	17	117	135	179	188	155	M	18	39	45	91	67	M	18	73	74	78	75
3	M	15	132	140	60		111	M	16	95	146	158	133	M	17	29	87	51	56
4	F	21	75	95	125		98	F	17	101	169	71	114	F	21	151	117	133	134
5	F	24	145	188	204	277	204	F	22	52	122	147	107	F	21	46	34	72	51
6	M	22	89	132	134	183	135	M	21	92	116	123	110	M	24	20	47	79	49
7	F	12	159				159	F	14	113	109	126	116	F	14	42	127	23	64
8	M	1½	171				171	M	7/12	154	194		174	M	2	28	93	96	72
9	F	10/12	91				91	F	5	137	111		124	F	1½	124	87	106	106
10	M	5	150	164	295		203	M	7	117	100	158	158	M	5	40	59	69	56
Over-all mean score							142						132						74

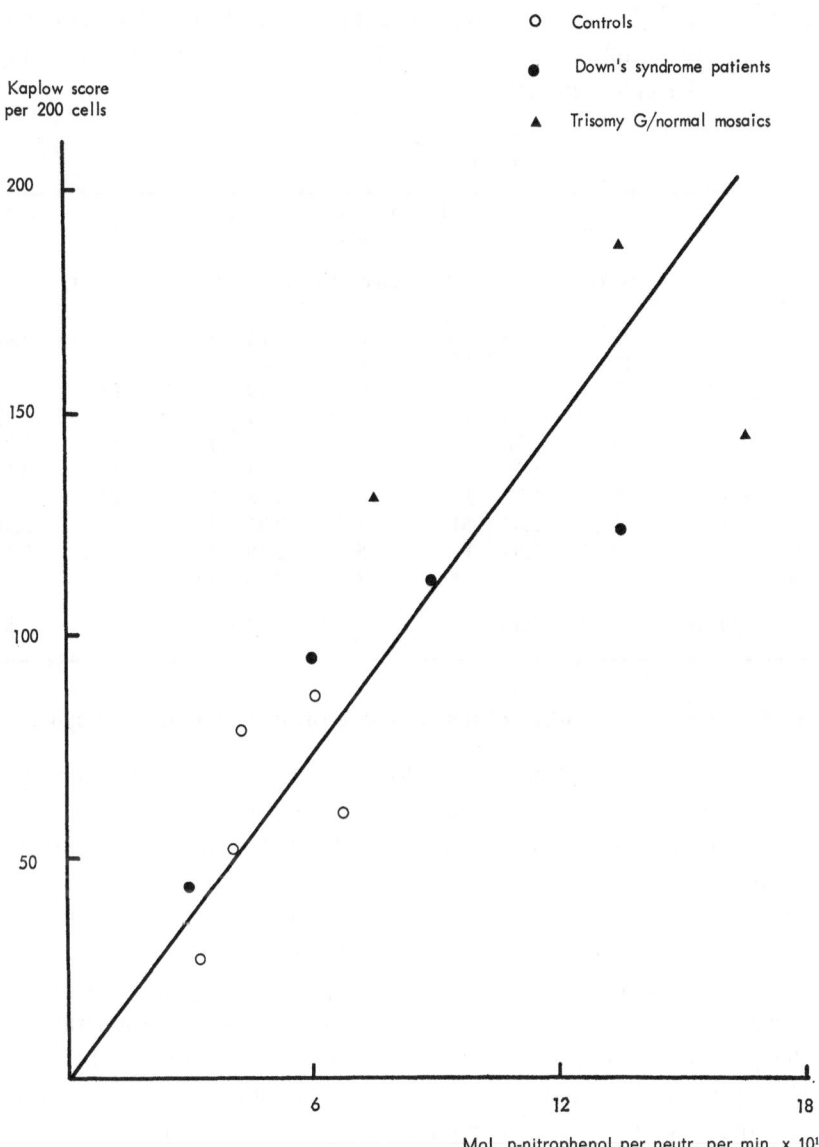

Fig. 16. Histochemical leukocyte alkaline phosphatase scores (Kaplow) plotted against results of biochemical determinations.

2. Lobe counts

Lobe counts in patients with Down's syndrome are lowered (Turpin and Bernyer 1947). The mean lobe counts of our controls, trisomy G/normal mosaics, and Down's syndrome patients, were 2.81, 2.56, and 2.34, respectively (Table XVII). No relationship was found between lobe counts and Kaplow indices.

Table XVII. Lobe counts

Patient No.	Trisomy G mosaics			Down's syndrome patients			Controls		
	Sex	Age (yr)		Sex	Age (yr)		Sex	Age (yr)	
1	F	25	2.54	F	18	2.04	F	25	2.40
2	M	17	2.38	M	18	2.42	M	18	2.77
3	M	15	2.68	M	16	2.19	M	17	2.89
4	F	21	2.90	F	17	2.39	F	21	3.15
5	F	24	2.38	F	22	1.99	F	21	2.92
6	M	22	1.78	M	21	1.93	M	24	2.50
7	F	12	2.77	F	14	3.08	M	14	3.08
8	M	1½	2.64	M	7/12	2.96	M	2	2.98
9	F	10/12	3.03	F	5	2.38	F	1½	2.74
10	M	5	2.49	M	7	2.06	M	4	2.71
	Mean:		2.56			2.34			2.81

3. Fine-structural studies of the granules of neutrophil granulocytes

The granules of neutrophil granulocytes have shown heterogeneity in morphological, cytochemical, and biochemical investigations.

The morphological heterogeneity has been demonstrated in detail by Bainton and Farquhar (1966, 1968), who differentiated in rabbit bone marrow between large, dense, azurophil, early formed and smaller, less dense specific granules which appeared in more mature cells. Azurophil and specific granules were also demonstrated in neutrophil granulocytes of guinea pigs by Brederoo and Daems (1970). In human peripheral blood the presence of four types of granules has been described by Daems (1968): round or oval electron-dense granules with a crystalline inclusion, less dense granules containing filaments, larger granules with coarse, floccular material, and smaller granules with fine floccular material. The author assumed a possible relationship between the first two types.

Cytochemical investigations were carried out by Ackerman (1964). Alkaline phosphatase activity in human bone marrow films appeared to reside within the specific granules. Bainton and Farquhar (1968) could in myelocytes of rabbit bone marrow also demonstrate alkaline phosphatase activity

62

corresponding to the distribution of the specific granules in agreement with the results obtained by Wetzel et al. (1967) and Spicer et al. (1968). Biochemical investigations have shown the lysosomal nature of granule fractions isolated from rabbit neutrophil granulocytes by their rich content of acid hydrolases (Cohn and Hirsch 1960). Baggiolini et al. (1969, 1970) differentiated between groups of granules in rabbit neutrophil granulocytes by zonal sedimentation and demonstrated that ninety per cent of the alkaline phosphatase content of neutrophils was located in the group of smaller, less dense, specific granules.

This was therefore in agreement with the results obtained in the cytochemical investigations.

The rise of the leukocyte alkaline phosphatase levels in trisomy G mosaics (van Gelderen et al. 1967) and Down's syndrome patients (Trubowitz et al. 1962) could either be related to changes in the numbers of different granules or result in an increase of the amount of enzyme per granule. A combination of these possibilities could occur and furthermore a different population of granules might be present in the neutrophil granulocytes of these patients, who also have a low lobe count of nuclei of neutrophil granulocytes.

To investigate the possibility of an increase of a certain kind of granule related to an increase of the alkaline phosphatase index, which would identify the localization of this enzyme, we studied the granules of the neutrophil granulocytes of four Down's syndrome patients, four trisomy G/normal mosaics, and three control patients.

a. Methods

To 1 ml of heparin (400 U/ml), 15 ml venous blood was added. This was divided into 2 portions, each of which was added to 2 ml 5 per cent dextran solution (mol.weight 200,000) and allowed to sedimentate for 30 minutes at 4°C. The plasma was then pipetted off and centrifuged for 10 minutes at 4°C at 900 x g, after which the supernatant fluid was pipetted off and the remaining cells were washed with physiological saline solution and centrifuged again at 4°C at 900 x g for 10 minutes. The supernatant fluid was aspirated and the washing with physiological saline and centrifuging repeated once. After the supernatant had been discarded, 1 per cent osmium tetroxide phosphate buffered to pH 3 was poured on to the sedimentated cells for a fixation period of 30 minutes at 4°C. The cells were then dehydrated in a graded alcohol series and

63

embedded in Epon 812. Ultrathin sections were cut on a Reichert ultramicrotome II, stained with uranyl acetate for 20 minutes followed by 10 minutes lead hydroxide staining. The sections were studied and photographed in a Siemens Elmiskop I or a Philips E.M. 200.

b. Results
The morphological heterogeneity described for the granules of normal human neutrophil granulocytes was confirmed by our observations. Three types of granules were differentiated: electron-dense (black) granules of varying sizes but mostly large, and less electron-dense (grey) granules of two sizes, large and small (Fig. 17). (The electron-dense granules of the first group correspond with the type 1 and type 2 granules described by Daems (1968), the big, less electron-dense granules with his type 3, the small, less electron-dense granules with type 4). The granules were counted in 10- to 20-cell sections of each patient. The results are shown in Table XVIII.

The same types of granules were found in the trisomy G/normal mosaics, Down's syndrome patients, and the controls. The total quantity of granules and the division into three types was fairly constant in the three groups. The means of the total numbers of granules and of the three different types show no significant differences between the three groups of patients. For the black electron-dense granules and the large and small less electron-dense grey granules $P \cong 0.25$: > 0.50: > 0.20, respectively (analysis of variance).

In addition, no relationship could be demonstrated between the Kaplow index and the total number or each of the three types of granules. The increased Kaplow index was therefore not reflected in an increase of either type or of the total number of granules, or by the appearance of a previously unknown type of granule. These results suggest an increase of the enzyme content within certain granules.

It may be said on the basis of these results that the leukocytes of the trisomy G/normal mosaic patients showed a significant increase of alkaline phosphatase levels as compared with controls. The increase was even higher, though not significantly, than that of Down's syndrome patients. Fine-structural studies failed to demonstrate the localization of alkaline phosphatase by qualitative or quantitative variations of granule populations of neutrophil granulocytes.

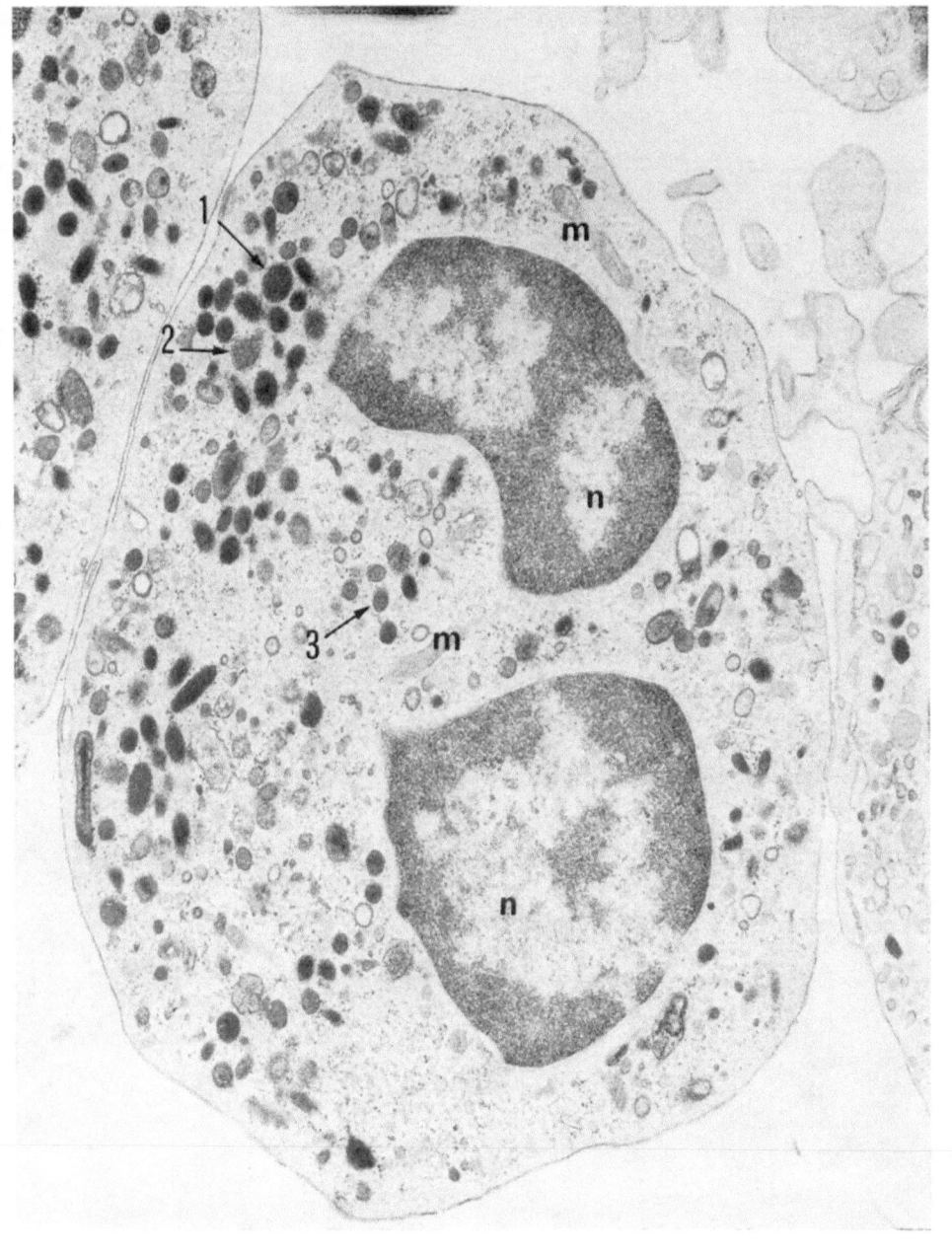

Fig. 17. Section of neutrophil granulocyte.

n = nucleus   1 = black granule   3 = small grey granule
m = mitochondrion   2 = large grey granule

Table XVIII. Mean granule count in leukocyte sections of trisomy G/normal mosaic patients, Down's syndrome patients, and controls

	Leukocytes	Black granules		Large grey granules		Small grey granules		Kaplow Index
	N	X	S	X	S	X	S	
Trisomy G/normal mosaics	12	50.9	17.12	25.8	10.17	84.8	22.63	91
	10	54.5	16.21	20.7	15.99	92.6	19.14	165
	10	38.1	12.91	26.8	12.01	130.3	29.64	196
	10	32.5	12.21	15.4	7.53	88.4	31.18	179
Down's syndrome patients	10	40.3	19.98	17.5	10.54	110.5	31.84	114
	10	63.7	25.85	23.8	11.18	79.5	26.04	101
	21	65.4	23.96	17.1	10.00	115.3	34.78	145
	16	50.8	23.10	22.8	9.77	123.3	38.64	52
Controls	18	46.0	19.68	22.3	9.27	121.9	40.45	99
	21	45.5	18.44	16.7	6.67	120.8	31.92	42
	10	56.7	10.14	19.7	5.42	120.8	27.71	23

N = Number of sections counted

X = Mean count

S = Standard deviation

65

The nuclear lobe counts of the neutrophil granulocytes of the trisomy G/normal mosaic patients showed intermediate values between the low lobe counts of patients with Down's syndrome and those of controls. A relationship between the lobe count and the Kaplow score of single granulocytes could not be detected.

The elevation of leukocyte alkaline phosphatase levels in trisomy G/normal mosaic patients and patients with Down's syndrome might be one of the results of a general disturbance of leukocytes in these patients, since various other enzymes also show increased activity, and a greater incidence of acute childhood leukaemia has been demonstrated by Krivit and Good (1956) and later confirmed by various other authors (Ager et al. 1956, Conen and Erkman 1966).

C. DISCUSSION OF THE RESULTS
OF THE VARIOUS INVESTIGATIONS

Duplications of chromosomes may lead to understanding of gene dosage effects of triallelic controlled characters and of deviation from frequency distributions of definite phenotypes, as stated by Bender et al. (1967). These effects of duplication would be expected to occur in mosaic patients in a degree related to their degree of mosaicism.

The occurrence of the investigated symptoms in our group of trisomy G/normal mosaic patients was compared with their occurrence in patients with Down's syndrome and controls. Individual mosaic patients were studied with respect to the relationship between the percentage of trisomic cells and the investigated symptoms. The results of the various investigations and the phenotypical manifestation of the extra G chromosome will be treated in this chapter.

I. CHROMOSOME STUDIES

The chromosome analyses of the trisomy G/normal mosaic patients show a high percentage of aneuploid cells in addition to the trisomy G. We also found a high percentage of aneuploid cells in patients with Down's syndrome. This is in agreement with the findings of Gall et al. (1970) who found 10 per cent hypomodal cells in 1158 cells with trisomy G. The percentage of aneuploid cells in normals varies in our laboratory from 5 to 10 per cent. It is possible that in patients with trisomic cells there is a general tendency toward irregular cell division that results in the high percentage of aneuploidy.

The occurrence of trisomy G/normal mosaics with a low percentage of euploid cells cannot be caused by random loss of G chromosomes, since a sub-population of these mosaics does not exist in any significant percentage of mongoloids, as had been pointed out by Gall et al. (1970). These authors found the loss of G chromosomes in mongoloids to be of the same order as in normal persons, where hypoploidy was shown by Neurath et al. (1970) to be caused by small chromosomal size.

The ratio between the percentage of trisomic cells in leukocytes and fibroblasts varied in our patients, contrary to the findings of Penrose and Smith (1966) who reported a twice higher frequency of trisomic cells in leukocytes. Patient 3 had 92 per cent trisomic fibroblasts, explaining his many mongoloid stigmata which were initially surprising when only 13 per cent of his leukocytes were known to be trisomic.

Insight into the phenotypical manifestations of the extra G chromosome is complicated by the varying degree of mosaicism in different tissues, which may cause a seeming discrepancy between the percentage of trisomic cells in the investigated tissue and the deviation toward Down's syndrome symptoms in the patient.

Taylor (1968, 1970) has described cell selection *in vivo* in trisomy G/normal mosaics. A rapid selection of either normal or trisomic cells occurred in the first two to three years of life of mosaic mongols. However, in all patients who ultimately had less than 10 per cent of either normal or trisomic leukocytes, this percentage was reached before the age of 12 months. Patients with more than 9 per cent normal or trisomic leukocytes at 12 months of age can therefore be expected to remain mosaic patients. One of our patients was only 9 months old, but had 17.5 per cent normal and 80 per cent trisomic cells and can be expected to have the required minimum of 9 per cent euploid cells at the age of 12 months to justify the diagnosis trisomy G/normal mosaicism, according to Penrose and Smith (1966). All the other patients satisfied these requirements.

II. CLINICAL EXAMINATIONS

The trisomy G/normal mosaic patients described in the literature were in all but a few cases selected on the basis of obvious or dubious signs of Down's syndrome. This selection and also the lack of uniformity in the descriptions of the patients makes it difficult to reach general conclusions about clinical signs in trisomy G mosaicism.

It is obvious that the intelligence quotient of our group of patients is influenced by their selection and therefore cannot be considered representative for trisomy G mosaicism. Five of our patients, for instance, were selected on account of a dubious or obvious diagnosis of Down's syndrome, while the other five patients were selected partially on account of mental retardation. This selection excluded *a priori* the inclusion of patients with normal intelligence quotients like those described by Clarke et al. (1963), Hayashi et al. (1962), Lindsten et al. (1962), and Aarskog (1969).

1. Body measurements

The body height, ratio upper/lower segment, and head circumference lay within Down's syndrome range for all the trisomy G/normal mosaic patients. This means values below the 10 percentile curve for body height for normal Dutch boys and girls.

It is not possible to find a relationship between percentages of trisomic cells and these measurements in individual patients, because the ages of the trisomy G/normal patients vary greatly.

2. Characteristic non-measurable signs of Down's syndrome

Ten phenotypical non-measurable signs of Down's syndrome were selected for incidence studies because of their power to discriminate between a normal individual and a Down's syndrome patient. The selected signs were: epicanthus, Brushfield's spots, abnormal ears, a furrowed tongue, abnormal teeth, a short neck, a curved 5th finger, a plantar furrow, cutis marmorata, and scant body hair. Several clinical signs considered important in this respect by other authors were for various reasons omitted from our investigations. Oblique palpebral fissures were not investigated in our patients because they are not easy to diagnose, and furthermore they were also found to occur in 37.5 per cent of a Swedish group with normal karyograms (Gustavson 1964). A highly arched palate is one of Øster's (1953) ten most characteristic signs, but Shapiro et al. (1967) found the palates of 153 Down's syndrome patients to be no higher than those of 1322 controls: the palates of the Down's syndrome patients were narrower and dramatically shorter, thereby giving the impression of greater height. A flat occiput is expressed in the measurements of skull circumferences and therefore in our opinion need not be stated separately. We do not refer to the occurrence of four finger creases, since they are far less specific for Down's syndrome than dermatoglyphs and also

69

occurred in 43.8 per cent of patients with normal karyograms (Gustavson 1964).

Our investigations have demonstrated the importance of taking age into account for the calculation of the incidence of certain clinical signs, since the two most frequently occurring signs of Down's syndrome in our series, abnormal teeth and a furrowed tongue, are not present in children younger than 1 to 2 years. On the other hand, epicanthic folds are also present in 38 per cent of the normal children under 5 years and therefore cannot serve as a diagnostic sign of Down's syndrome in young children.

Our control group has been regarded under the restriction of their age, but it nevertheless gives a good impression of the incidence of most non-measurable clinical signs at all ages. Since 27 of the 80 control children did not have a single sign of Down's syndroma, 31 had one, ten had two, twelve had three positive signs, and none had more than three signs, the presence of three or more positive non-measurable signs seems to justify further investigation. The range of occurrence of positive signs in the patients with Down's syndrome was 2-10, with a mean value of 5.3.

Our clinical results demonstrate the presence of two to eight non-measurable signs of Down's syndrome in our trisomy G/normal mosaic patients, seven of the ten mosaic patients having four or more of these ten signs.

Although the average frequency of non-measurable clinical signs in our trisomy G/normal sample is lower than that in Down's syndrome, the difference is too small for practical use in differentiation. No relation was found between the number of positive signs and percentages of trisomic cells in individual trisomy G/normal mosaic patients.

The clinical diagnosis of Down's syndrome or trisomy G/normal mosaicism in newborns is of special importance, and may be difficult even in the hospital. Zapella and Cowie (1962) reported that the diagnosis had only been accurately made in 92 out of their 141 cases. The presence of Brushfield's spots, plantar furrows, and marked cutis marmorata should be added to the positive signs described by Hall (1964) in newborns, since they are easy to diagnose and seldom occur in normals.

Since the selected clinical symptoms, especially in the case of mosaicism, may be insufficient to make diagnosis possible, additional information is required. Dermatoglyphs and pelvic measurements should be analysed. The maternal age at the birth of the patient, the birth weight and mental development, and finally leukocyte alkaline phosphatase levels can be studied. All these investigations take relatively little time compared to the time needed for karyotyping.

3. Dermatoglyphs

The dermatoglyphs of patients with doubtful Down's syndrome can be of great help in establishing the diagnosis, since all patients with a percentage of trisomic cells above 30 in either skin or blood have had positive log indices (Ford Walker), and the occurrence of single dermatoglyphic traits of Down's syndrome in patients with negative log indices also has value. The dermatoglyphs of the parents of our trisomy G mosaic patients did not show the resemblance to Down's syndrome patterns stated by Penrose (1954) for parents of children with Down's syndrome.

4. Pelvic measurements

Especially in newborns where there is uncertainty about the diagnosis of Down's syndrome, pelvic measurements are of great value. No data on pelvic angles in older children were found in the literature. The data collected by van Gelderen (1967) showed no difference between the acetabular angles of children with Down's syndrome and those of controls older than 4 years, but the difference between iliac angles remained about the same as in newborns. It therefore seems that only iliac angles are of diagnostic value in older patients, where they will in any case often be the only measurable angles, since acetabular angles cannot be measured after the calcification of the Y cartilages.

In our patients the diagnostic value of pelvic measurements was illustrated by the fact that five of the eight investigated mosaic patients fell within the range of Down's syndrome, whereas only two of the eight controls lay in the overlapping zone of Down's syndrome and control range. A relationship between the percentage of trisomic cells and iliac angles is indicated by the fact that the two patients with the lowest percentages of trisomic cells are within control range and the two patients with the highest percentages of trisomic cells are within the range of patients with Down's syndrome. Patient 4, who also has the Rubinstein-Taybi syndrome, has the small pelvic angles of this syndrome described by Taybi and Kane (1968). Differentiation between trisomy G/normal mosaic

71

patients and patients with Down's syndrome is not possible on the basis of pelvic measurements.

III. MATERNAL AGES AND BIRTH WEIGHTS

The mean maternal age at birth of our trisomy G/normal mosaic patients was 33.1 years. Unlike Richards (1969), we found no significant differences in mean maternal ages between the groups of Down's syndrome patients and trisomy G/normal mosaics. However, his data were compiled from the entire literature and pertained to populations with varying mean maternal age. They have little value in this respect. Richards (1969) thought that the lower mean ages of mosaic patients compared with patients with Down's syndrome were caused by the origination of a number of these patients from normal zygotes of younger mothers. Our patient material is of course small, but it supplies no support for this theory.

The mean birth weight of our trisomy G/normal mosaic patients was 2993 g, even slightly below the mean birth weights of 3141 g and 3035 g found by Swaak (1967) for 125 boys and girls with Down's syndrome, respectively. These mean birth weights were significantly lower than the mean birth weights for normal Dutch infants. In our study we could not differentiate between intra-uterine growth retardation and increased frequency of prematurity, as already mentioned.

IV. LEUKOCYTES

1. Alkaline phosphatase

Alkaline phosphatase scores were determined histochemically according to Kaplow (1955) and biochemically. The results of both methods were in good agreement, thereby demonstrating the usefulness of the Kaplow method.

If there is a direct relationship between the extra G chromosome and the leukocyte alkaline phosphatase index, a relationship between the percentage of trisomic cells of trisomy G mosaics and the Kaplow index would be expected in the first place. In the second place the over-all mean score of a group of trisomy G/normal mosaic patients would then lie between those of Down's syndrome patients and normals. No relationship, however, existed in our investigations between the Kaplow index and the number of trisomic cells. This is contrary to De Carli's (1964, 1968) observation, who found a correlation between enzyme levels and the number of small acrocentric chromosomes in cell lines derived from human embryonic epithelium. The over-all mean Kaplow index of the trisomy G/normal mosaics was even higher than that of the Down's syndrome patients. The difference between these groups was not significant, but they were both significantly higher than the control group. The diagnostic value of the Kaplow score in individual patients is demonstrated by the elevated levels in 7 out of 10 mosaic patients and 9 out of 10 patients with Down's syndrome, whereas only 2 controls had scores above 100.

In favour of an indirect instead of a direct relationship between leukocyte alkaline phosphatase enzyme levels and the extra G chromosome, are the following findings. In Down's syndrome patients other leukocyte enzymes have been found to be elevated as well, such as acid phosphatase, (Mellman et al. 1964), the X-linked glucose-6-phosphate dehydrogenase and galactose-1-phosphate uridyl transferase (Brandt et al. 1963, Hsia et al. 1964, Kelly et al. 1967). Finally, leukocyte alkaline phosphatase does not always parallel the presence of a Philadelphia chromosome. Lowered levels have been found when a Ph.-chromosome was absent (Block et al. 1963), and elevated alkaline phosphatase levels could occur in the presence of a Ph.-chromosome (Teplitz et al. 1964). The fact that leukocyte alkaline phosphatase is related to leukocyte abnormality is indicated by the finding of normal levels of this enzyme in fibroblasts (Cox 1965, Nadler et al. 1967).

2. Lobe counts

The lobe counts of leukocyte nuclei showed lowered values in Down's syndrome patients; intermediate values were found in the trisomy G/normal mosaic patients. This led to investigations of leukocyte kinetics which have given conflicting results. An increase of leukocyte turnover was found by Raab et al. (1966) and Mellman et al. (1968), whereas Pearson (1967) and Galbraith and Valburg (1966) found a normal turnover.

No correlation was found in our study between lobe count and Kaplow index, contrary to the findings of Rosen and Nishiyama (1968), who concluded that segmented ('older') neutrophils contribute a major share of leukocyte alkaline phosphatase activity.

3. Fine-structural studies

The electron-microscopical investigations, which showed the occurrence of the same population of granules of neutrophil granulocytes with no differences in divisions or in total amounts of granules between trisomy G mosaics, Down's syndrome patients, and controls, provided no information about the localization of alkaline phosphatase in any type of granules in these morphological studies, but did show that the morphology and number of the three types of granules is not affected by the extra G chromosome.

V. CONCLUSION

Our studies on trisomy G/normal mosaic patients were undertaken in an attempt to detect the phenotypical manifestation of the extra G chromosome. If there were a direct relationship between this extra chromosome and a measurable criterion of Down's syndrome, intermediate values related to the degree of mosaicism could be expected to occur in the trisomy G/normal mosaic patients, and non-measurable criteria of Down's syndrome might be expected to occur in a lower percentage of trisomy G/normal mosaic patients.

The various investigations have demonstrated wide phenotypical variations in the trisomy G/normal mosaic patients. The patients range from mentally retarded individuals with atypical Down's syndrome to patients with all the signs of this syndrome.

Body-measurement values lay within the Down's syndrome range for all trisomy G/normal mosaic patients. On careful clinical inspection, some of the non-measurable characteristic signs of Down's syndrome were found to be present in all the mosaic patients in a significantly higher incidence than in controls. The total number of non-measurable signs in individual patients was not directly related to the percentage of trisomic cells, but the mean for these signs in the group was intermediate between that for Down's syndrome patients and controls. Dermatoglyphic log indices provided an indication of the degree of mosaicism, because they were only positive in patients with more than 30 per cent trisomic cells in either skin or blood. Iliac angles also deviated more in the direction of the values of Down's syndrome in patients with higher percentages of trisomic cells. Mean leukocyte alkaline phosphatase levels were even higher in the group of mosaic patients than in the group of Down's syndrome patients.

It is impossible to differentiate on clinical grounds between trisomy G/normal mosaic patients with high percentages of trisomic cells and patients with Down's syndrome. On the other hand, it is of even more interest to detect trisomy G/normal patients with atypical Down's syndrome or low percentages of trisomic cells, which should be possible once some of the described non-measurable clinical signs have been detected in these patients and supplemented by dermatoglyphic, alkaline phosphatase, and pelvic measurement data. All these parameters together will provide an indication for the performance of chromosome studies with much more chance of revealing an abnormality than would random culturing of cells of patients.

SUMMARY

Down's syndrome patients have an extra G chromosome in all their cells, whereas the trisomy G/normal mosaic patients have this extra chromosome only in some of their cells. For this reason, a direct relationship between any measurable sign of Down's syndrome and the extra G chromosome could be expected to result in an occurrence of that sign in mosaic patients in a degree proportional to the percentage of trisomic cells, while non-measurable clinical signs would only appear in a percentage of mosaic patients.

Our investigations, which included chromosomal, clinical, and leukocyte studies in ten trisomy G/normal mosaic patients, are reported, and the literature on trisomy G/normal mosaicism is reviewed. The patients were compared with normal controls and Down's syndrome patients. Five of the ten mosaic patients were selected for karyotyping because no explanation could be found for their mental and growth retardation combined with other congenital anomalies; the other five mosaic patients had dubious signs of Down's syndrome.

The results of the chromosome studies of blood and skin showed a varying ratio of trisomic cells in both. Clinical investigations of body measurements and non-measurable signs of Down's syndrome, with reference to the literature on this subject, have demonstrated values within Down's syndrome range for the measurable signs of all mosaic patients. Some of the non-measurable signs of Down's syndrome occurred in all the mosaic patients, but were absent in more than 30 per cent of control children. The importance of selection of clinical signs according to age is stressed, especially in relation to newborns. The presence of more than three of the ten selected clinical signs is highly suggestive of either trisomy G/normal mosaicism or Down's syndrome. Dermatoglyphic log indices (Ford Walker) were positive in all patients with more than 30 per cent trisomic cells in either skin or blood, and negative in patients with less than 30 per cent trisomic cells in either skin or blood. Pelvic measurements were within the mongoloid range for five of the eight patients in whom the pelvis was X-rayed. One patient had the symptoms of the

Rubinstein-Taybi syndrome as well as the characteristic dermatoglyphs and pelvic measurements of this syndrome.

The leukocyte investigations included histochemical and biochemical alkaline phosphatase determinations, and fine-structural studies of granules of neutrophil granulocytes were carried out. Mean alkaline phosphatase levels of trisomy G mosaic and Down's syndrome patients were significantly higher than the mean level of controls (P < 0.01) but did not differ significantly mutually (P > 0.50).

The mean lobe count of trisomy G mosaics was intermediate between those of Down's syndrome patients and controls. No relationship could be found in individual patients between alkaline phophatase levels and percentages of trisomic cells, which seems to support the hypothesis of a more general disturbance of leukocytes in Down's syndrome patients. Fine-structural studies revealed corresponding populations of granules of neutrophil granulocytes in the mosaics, Down's syndrome patients, and controls, with no significant differences between total amounts or division into certain types (P values > 0.20). These studies were therefore unable to provide an indication of the localization of alkaline phosphatase activity.

The results of the various investigations are discussed. They indicate that careful clinical investigations may lead to a possible diagnosis of either Down's syndrome or trisomy G mosaicism in patients in whom this diagnosis was not thought of initially. Patients with more than 30 per cent trisomic cells will also have a positive dermatoglyphic log index and possibly lower iliac angles. It will be impossible to differentiate between these mosaic patients and patients with Down's syndrome on clinical grounds, since no linear relation could be found between the positive signs of Down's syndrome and the percentages of trisomic cells.

The results of the study as a whole support the assumption that the extra G chromosome in our cases of mosaicism is the same as that found in Down's syndrome.

CASE HISTORIES

Patient 1, F. (22.5.1944)
This patient is one of a twinpair born after a full-term uncomplicated pregnancy with a birth weight of 1500 g; the twin brother, who weighed 4000 g at birth, and the other 9 brothers and sisters are all healthy. No mental or other illnesses are known in the family. The maternal age at the patient's birth was 32 years. The patient had feeding problems and developed slowly. From the age of one year she often had otitis media with perforation of the ear drums. Her motor development was retarded as compared to that of the twin brother; she started to walk at the age of 2. When she was 13, she was hospitalized for acute glomerulonephritis, and a mastoid operation was performed. Menstruation started at the age of 15. At 24, she has developed into a friendly girl, who can do simple work and talks reasonably well. The I.Q. is 51. Her height is 157.5 cm with a U/LS index of 1.03. The head circumference is 55 cm, the palpebral fissures are oblique, and she has strabismus but shows neither epicanthus nor Brushfield's spots. The ears are abnormal. She has large, irregular, protruding teeth in the upper jaw; the tongue is large and partly furrowed. Heart, lungs, and abdomen show no abnormalities. The breasts are well developed, pubic hair is sparse, there is no cutis marmorata. The fingers are short and broad, muscular tonus is rather high, reflexes are normal. The dermatoglyphic index (Ford Walker) is — 4.93. Radiological investigations gave a mean iliac angle of 71°, i.e. within the normal range. The Kaplow index is 97.

Chromosome studies were undertaken on the basis of mental and growth retardation for which no exogenous reasons could be found. Leukocyte and fibroblast cultures showed 7 and 22 per cent trisomic cells, respectively.

Patient 2, M. (26.5.1951)
The patient was delivered normally after an uneventful pregnancy as the tenth child of a 43 year old mother. His birth weight was 3000 g. Between the ages of 2 weeks and 3 months he was hospitalized on

account of otitis and frequent convulsions. He learned to walk at the age of 18 months but has never talked. At the age of 4 years he was placed in the above-mentioned institution because of the convulsions. There he developed, after initial feeding problems, into a sturdy looking boy.

At the age of 19 years the patient has a height of 154 cm. The upper/ lower segment quotient is 1.03. His general appearance does not give the impression of Down's syndrome, although he has the following stigmata: epicanthus, large, irregularly implanted teeth, a high triradius on both hands, and curved 5th fingers. The head circumference is 51 cm and has been small from birth, e.g. 43 cm at 10 months. There are no anomalies of the heart, lungs, or abdominal organs, and his testes are normally developed. His I.Q. is estimated to be below 40. His brothers and sisters are normal; there is no history of mental illnesses in the family.

The percentage of trisomic cells in two leukocyte and one fibroblast culture were 12, 6 and 15, respectively.

Patient 3, M. (3.3.1953)
The patient was born 3 weeks early from a 32 year old mother after a normal pregnancy. The birth weight was 2700 g. He was operated on a few hours after birth because of an imperforate anus. His temperature was below normal during infancy. He developed slowly, starting to sit at the age of 3 and to walk at the age of 4½. Between his 1st and 5th year he suffered from infections of the urinary tract. Retention of the left testis was corrected at the age of 7. At the age of 12 an arthrodesis was carried out on both feet to correct flat feet.

At the age of 16 the patient is a quiet, cooperative boy who cannot talk, with an I.Q. estimated at about 40. His height is 142.5 cm with an U/LS index of 0.98. The head circumference is 51 cm with a flattened occiput. The eyes show no epicanthus, but Brushfield's spots are present. Nose and ears have a normal shape. The tongue is neither large nor furrowed. Heart and lungs show no abnormalities. There is diastasis of the abdominal rectal muscles. The left testis is present, the right absent. Pubic hair has grown.

The extremities show acrocyanosis of the hands and feet. His 5th fingers are short and curved, his feet show a plantar furrow, and the 2nd and 3rd, and the 4th and 5th toes are grown together. There is no cutis marmorata; the tonus of trunk and upper extremities is low, that of the legs high. Reflexes are normal.

He has an older and a younger brother, both of whom are healthy; no mental illnesses are known in the family.

His dermatoglyphic index (Ford Walker) is + 3.99, leukocyte alkaline phosphatase (Kaplow): 111.

Radiological investigation showed a mean iliac angle of 52.5°, i.e. approaching the mean angle of Down's syndrome patients (47.7°). His EEG showed signs of primary generalized epilepsy.

Patient 4, F. (8.7.1947)

The patient was born from a 27 year old mother by forcipal extraction after a full-term pregnancy. The birth weight was 4000 g. The mother suffered from tuberculosis, for which a tomogram of her lungs was made when she was 6½ months pregnant. The patient is the third of 5 children; the other 4 are healthy. A cousin of the mother's is mentally retarded but not a case of Down's syndrome; the other members of the big family are normal.

The patient developed slowly. She had feeding problems and learned to walk at the age of 2. At 21 she has developed into a chubby looking, short girl who is mentally retarded and has all the clinical characteristics of the syndrome first described by Rubinstein and Taybi (1963) and bearing their names. She has the characteristic flattening of the distal phalanxes of toes and thumbs. Her face shows the antimongoloid slant of the eyes, with an epicanthus and a slight strabismus; the nose is beaked with a low septum nasi, the palate highly arched. The ears are set low with a slightly deformed helix. The head circumference is 50 cm. The heart, lungs, and abdominal organs are normal. The breasts are well developed, pubic hair is present, and she menstruates. Tonus and reflexes are normal. Her height is 142.5 cm, with a U/LS index of 1.16. The I.Q. is estimated at about 40.

The dermatoglyphs (Fig. 12, Table IX) have a log index (Ford Walker) of — 4.42. The characteristic pattern of the Rubinstein-Taybi syndrome in the first interdigital area (Giroux 1967) is present. Both palmar triradii are distantly placed. An additional proximal loop in the hallucal area, as described by Berg (1966), is not present.

Radiograms show the broad terminal phalanxes of the thumbs and big toes. Her mean iliac angle is 45.5°, well below the mean for normals, in agreement with the low iliac angles found in 4 of these patients by Taybi and Kane (1968).

Two chromosome studies were done of leukocytes and one of fibroblasts. The percentages of trisomic cells were 13, 14, and 3, respectively. The leukocyte alkaline phosphatase index (Kaplow) was 98, i.e. slightly elevated.

Patient 5, F. (13.7.1944)

The patient was delivered normally after a full-term pregnancy. The birth weight was 3750 g. She had feeding problems at the age of 3 months. Her development was retarded; she learned to walk at the age of 6 years.

The patient is the ninth of 12 children; the others are healthy. The age of the mother at birth was 37 years. Except for diabetes and asthma, no chronic illnesses are reported for the family.

She has now grown into a severely retarded girl of 25, with an I.Q. below 40. Her height is 150 cm, her head circumference 50 cm. Her face is broader at the base, which, in combination with the close set eyes, gives her a slightly ape-like appearance. She has neither epicanthus nor Brushfield's spots. Nose, ears, mouth, and tongue are normal. She has a short neck. Heart, lungs, and abdominal organs show no abnormalities. She has tapering fingers with short 5th fingers. The breasts are small, pubic hair is present. There have been no menses, but the sex-chromatin is positive. Her tonus is high, and the reflexes are lively with a positive Babinsky in both feet. She moves with a stiff gait, walking on her toes.

The dermatoglyphs show fingertips with 8 loops, one arch, and one whorl. On the palms of both hands the triradius is situated distally and there are patterns in the 3rd interdigital area. Both hallucal patterns are whorls. The log index (Ford Walker) is – 1.01, a value which occurs in only 2 per cent of the Down's syndrome patients.

Radiological investigation showed a mean iliac angle of 58.5°. Leukocyte alkaline phosphatase is 204 (Kaplow). An extra chromosome of the G-group was found in 15 and 11 per cent of leukocyte and fibroblast cultures, respectively.

Patient 6, M. (26.7.1946)

The patient is the second child of a 32 year old mother who was kept on a salt-free diet during pregnancy because of albuminuria. The delivery lasted 14 hours, after which the baby was born drowsy and limp, crying weakly. His birth weight was 3500 g. His development was retarded. At the age of 22 he cannot talk, but is able to walk and dress himself. There is very little contact possible with the patient, whose I.Q. is estimated at about 40. He had one epileptic insult, but his EEG showed no signs of epilepsy. There were no other illnesses.

Physically, he is a small boy with a height of 159.5 cm and an U/LS index of 0.98. His head has a triangular shape with a flattened occiput and a circumference of 52 cm. The palpebral fissures are oblique; he has a strabismus convergens, but neither epicanthus nor Brushfield's spots are present. The upper part of the ears is flaring and bent outwards. The nose is narrow. He has irregular small teeth, a highly arched palate, and a slightly furrowed tongue. Heart, lungs, and abdominal organs are normal. The genitals are underdeveloped. The 5th fingers are curved; the feet are concave and the metatarsals, especially the left metatarsal IV, are very short. He walks on the medial side of his feet with a stiff gait, taking small steps. His tonus and reflexes show no pathological changes. There is no cutis marmorata; his body hair is normal. His I.Q. is estimated below 40.

An older brother and a younger brother and sister are healthy. No mental illnesses are known to exist in the family.

His dermatoglyphs have a log index (Ford Walker) of + 4.32, the Kaplow index is 135. Radiological investigations of the hands showed shortening and deformation of the middle phalanx of digit V and variations and growth retardation of the carpals characteristic of Down's syndrome. The feet show a spreading of the metatarsals and toes II-V, a shortening of the left metatarsal IV, and shortened middle phalanxes. His mean iliac angle is 55°, i.e. within the mongoloid range. The percentages of trisomic cells in blood and skin cultures are 30 and 50, respectively.

Patient 7, F. (19.7.1957)
The patient was delivered by forceps after a full-term uneventful pregnancy, as the second child of a 36 year old mother. The birth weight was 3200 g and the infant had a mild jaundice. She was hospitalized for 5 months, starting at the age of 4 months, because of feeding problems and growth retardation. She suffered from short periods of loss of consciousness. Her weight at the age of 12 months was only 4500 g. Development was retarded; she started to sit at 1½ years and to walk at the age of 2½. At the age of 12 years she was hospitalized for osteomyelitis of

the right clavicle. At the age of 12½ years her body height is 133.2 cm, which is under the 10th percentile for Dutch girls of this age. The upper/lower segment index is 0.98, the head circumference 49.6 cm. There are epicanthic folds at both eyes, a double helix is present on the left ear, the teeth are large and coarse, the tongue is furrowed. She does not have a short neck, plantar furrows, or cutis marmorata. The body hair is not sparse. Dermatoglyphs are well within the mongoloid range with a log index of + 7.57 (Ford Walker).

The patient can say a few words and is cooperative. Her I.Q. is estimated at about 50. Leukocyte alkaline phosphatase (Kaplow) is 159. No mental illnesses are known in the family.

Although the patient has several signs of Down's syndrome, she does not have the typical appearance of these patients. Her hair is curly, the face is narrow with a normal nose, and she has a slender figure and long fingers. Chromosomes were studied because of the atypical Down's syndrome. Karyotyping revealed 60 per cent trisomic and 34 per cent euploid cells.

Patient 8, M. (22.12.1967)
The patient was born as the first child of a 25 year old mother who was kept on a salt-free diet during the last 6 weeks of the pregnancy. The delivery was without complications. The child cried immediately, the birth weight was 2800 g.

He had feeding problems at 3 months, and his temperature was persistently below normal. At the age of 6 months his parents took him to the hospital because they suspected that he had Down's syndrome. No mental illnesses were known in the family. His length was then 61 cm, the head circumference 41 cm. He laughed at the age of 6 weeks. His face gave a mongoloid impression because of the flat nose, the oblique palpebral fissures, and the protruding tongue, which was not furrowed. Epicanthus and Brushfield's spots were present; the ears showed no abnormality. Heart and lungs were normal; a slight umbilical hernia was present. On both feet a plantar furrow could be seen between the big and 2nd toe. His hair was shiny; there was no cutis marmorata. His dermatoglyphs fell well within the mongoloid range, with a log index of + 2.89 (Ford Walker). His iliac index is 65, which at his age means a 90 per cent probability of Down's syndrome. Leukocyte alkaline phosphatase (Kaplow) was 171.

Three leukocyte cultures and one fibroblast culture were carried out, the percentages of trisomic cells being 75, 66, 80, and 78, respectively.

The patient has developed well so far. He suffered from an otitis with perforation once, but there were no other illnesses. He learned to walk at 16 months. At 2½ years he can speak a few words. His weight at that age is 14.1 kg, height 83 cm, and head circumference 47 cm, the retardation in height being the greatest. His dentition is completed with the exception of the canines and one lateral incisor in the lower jaw. His tongue is now furrowed.

Patient 9, F. (19.5.1969)

The patient, a 10 month old girl, was delivered normally after an uneventful, full-term pregnancy. The mother was then 41 years old. The birth weight was 3380 g. The half-brother and half-sister from the mother's first marriage and the half-brother from her second marriage are healthy. The patient's father, her mother's third husband who was born in Suriname, is healthy. There are no known mental illnesses in the family.

The patient has developed well so far. She has not been ill, but has frequent colds. She is a happy, rather fat, responsive child, with a marked ability for imitation. She is able to sit up with help. Her height is 67 cm. Her face gives a mongoloid impression because of the bilateral epicanthus and the shape of her nose. Other stigmata of Down's syndrome are Brushfield's spots, a short neck, plantar furrows, and cutis marmorata; a systolic murmer grade II and an umbilical hernia are also present.

The dermatoglyphs show a high triradius on the left hand, and an arcus tibialis is present on both feet.

However, the nose, mouth, tongue, and shape of the head are normal, there is no hypotonus, and her development is only slightly retarded. For these reasons the diagnosis of Down's syndrome was questioned and leukocyte cultures were performed. They showed a trisomy G in 80 per cent of the cells.

The leukocyte alkaline phosphatase level was 91 (Kaplow).

Patient 10, M. (26.5.1963)

The patient was born as the first child of a 26 year old mother. His birth weight was 2850 g, with a length of 45 cm. He drank slowly, and had occasional convulsions in his first year. An EEG made in 1966 showed abnormalities. Apart from otitis and frequent colds, for which his adenoids and tonsils were removed, he had no illnesses.

At 6 years he has developed into a small boy of 97 cm, with an U/LS index of 1.31, and a definitely mongoloid face. He can sit and walk, but does not talk. His I.Q. is estimated to be about 40. His head circumference is 47.5 cm, his eyes have Brushfield's spots, and there is epicanthus and strabismus. The nose is flattened, the tongue furrowed, the ears are of normal shape but implantation is low. Heart, lungs, and abdominal organs show no abnormalities. The right testicle is present, the left can be felt in the inguinal canal. There is no cutis marmorata; he has a general hypotonus and normal reflexes. His hands are plump and short, the dermatoglyphs have a log index of $+$ 6.83 (Ford Walker). Leukocyte alkaline phosphatase (Kaplow) is 203.

Radiological investigation showed an iliac angle of 44°, i.e. well within mongoloid range.

The mother had an extra-uterine pregnancy a year after the patient's birth, and two years later she gave birth to another son who is normal. His father has a cousin with Down's syndrome. No other cases of mental disease are known in the family.

Leukocyte and fibroblast cultures, carried out because of the young maternal age, showed percentages of trisomic cells of 84 and 85, respectively.

REFERENCES

Aarskog, D. (1969). Down's syndrome transmitted through maternal mosaicism. *Acta Paediat. Scand.* 58, 609.

Ackerman, G. A. (1964). Histochemical differentiation during neutrophil development and maturation. *Ann. N.Y. Acad. Sci.* 113, 537.

Ager, E. H., Schuman, L.M., Wallace, H. M., Rosenfield, A. B. and Gullen, W. H. (1965). An epidemiological study of childhood leukemia. *J. Chron. Dis.* 18, 113.

Alter. A. A., Lee, S. L. (1968). Leukocyte alkaline phosphatase in the 21-trisomy (Down's) syndrome. *Ann. N.Y. Acad. Sci.* 155, 1023.

Armendares, S., Urrusti-Sanz, J. and Diaz-del-Castillo, E. (1967). Iliac index in newborns. Comparative values at term, in prematurity, and in Down's syndrome. *Amer. J. Dis. Child.* 113, 229.

Astley, R. (1963). Chromosal abnormalities in childhood, with particular reference to Turner's syndrome and mongolism. *Brit. J. Radiol.* 36, 2.

Aula, P., Hjelt, L. and Kauhtio, J. (1961). Chromosal investigation in congenital malformations. *Ann. Paediat. Fenn.* 7, 206.

Baggiolini, M., Hirsch, J. G. and Duve, C. de (1969). Resolution of granules from rabbit heterophil leukocytes into distinct populations by zonal sedimentation. *J. Cell Biol.* 40, 529.

Baggiolini, M., Hirsch, J. G. and Duve, C. de (1970). Further biochemical and morphological studies of granule fractions from rabbit heterophil leukocytes. *J. Cell Biol.* 45, 586.

Baikie, A. G., Court Brown, W. M., Buckton, K. E., Harnden, D. G., Jacobs, P. A. and Tough, I. M. (1960). A possible specific chromosome abnormality in human chronic leukemia. *Nature* 188, 1165.

Bainton, D. F. and Farquhar, M. G. (1966). Origin of granules in polymorphonuclear leukocytes. *J. Cell Biol.* 28, 277.

Bainton, D. F. and Farquhar, M. G. (1968). Differences in enzyme content of azurophil and specific granules of polymorphonuclear leukocytes. *J. Cell Biol.* 39, (I), 286 (II), 299.

Barkla, D. H. (1966). Congenital abscence of permanent teeth in mongols. *J. Ment. Defic. Res.* 10,198.

Bartalos, M. and Baramki, T. A. (1967). *Medical cytogenetics.* The William and Wilkins Company, Baltimore.

Benda, C. E. (1969). *Down's syndrome. Mongolism and its management.* Grune and Stratton, New York, London.

Bender, K., Ritter, H. and Wolf, U. (1967). Zur Frage der Zuordnung von Genen zu bestimmten Autosomen des Menschen mit Hilfe von Chromosomen Aberrationen. *Humangenetik.* 4, 85.

Blank, C. E., Gemmell, E., Casey, M. D. and Lord, P.M. (1962). Mosaicism in a mother with a mongol child. *Brit. Med. J.* 2, 378.

Block, J. B., Carbone, P. P., Oppenheim, J. J. and Frei, E. (1963). The effect of treatment in patients with chronic myelogenous leukemia. Biochemical studies. *Ann. Int. Med.* 59, 629.

Brandt, N. J., Frøland, A., Mikkelsen, M., Nielsen, A. and Tolstrup, N. (1963). Galactosaemia locus and the Down's syndrome chromosome. *Lancet* 2, 700.

Brederoo, P. and Daems, W. Th. (1970). Submicroscopic cytology of guinea pig peritoneal exudates. I. The heterogeneity of the granules of neutrophilic granulocytes. *Microscopie Electronique 1970.* Résumés des communications présentées au septième congrès international, Grenoble, (P. Favard, ed Vol. III) p. 541. Société Française de Microscopie Electronique, Paris.

Brøgger, A. (1966). *Translocation in human chromosomes.* Universitetsforlaget, Oslo, Norway.

Caffey, J. and Ross, S. (1956). Mongolism (mongoloid deficiency) during early infancy. Some newly recognized diagnostic changes in the pelvic bones. *Pediatrics,* 17, 642.

Caffey, J. and Ross, S. (1958). Pelvic bones in infantile mongoloidism. Roentgenographic features. *Amer. J. Roentgenol.* 80, 458.

Chaudhuri, A., Chandra, R. K., Kaul, K. K., Dabke, A. T. and Chaudhuri, K. C. (1968). A possible case of trisomy 22. *J. Ment. Defic. Res.* 12, 177.

Chaudhuri, A. and Chaudhuri, K. C. (1965). Chromosome mosaicism in an Indian child with Down's syndrome. *J. Med. Genetics.* 2, 131.

Chitham, R. G. and Mac Iver, E. (1964-1965). A cytogenetic and statistical survey of 105 cases of mongolism. *Ann. Human Genet.* 28, 309.

Chu Ch'ang-Ning and Chung Lien-Yun (1966). Chromosomal mosaicism and Down's syndrome. *Chinese Med. J.* 85, 337.

Clarke, C. M., Edwards, J. H. and Smallpeice, V. (1961). Trisomy - normal mosaicism in an intelligent child with some mongoloid characters. *Lancet* 1, 1028.

Clarke C. M., Ford, C. E., Edwards, J. H. and Smallpeice, V. (1963). Trisomy - normal mosaicism in an intelligent child with some mongoloid characters. *Lancet* 2, 1229.

Cohen, M. M. and Winer R. A. (1965). Dental and facial characteristics in Down's syndrome (mongolism). *J. Dental Res.* 44, suppl. to no 1, 197.

Cohn, Z. A. and Hirsch, J. G. (1960). The isolation and properties of the specific cytoplasmic granules of rabbit polymorphonuclear leukocytes. *J. exp. Med.* 112, 983.

Conen, P. E. and Erkman, B. (1966). Combined mongolism and leukemia. *Amer. J. Dis. Child.* 112, 429.

Cox, R. P. (1965). Regulation of alkaline phosphatase in skin fibroblast cultures from patients with mongolism. *Exp. Cell Res.* 37, 690.

Cummins, H. (1936). Dermatoglyphic stigmata in mongoloid imbeciles. *Anat. Rec.* 64, suppl. 11.

Daems, W. Th. (1968). On the fine structure of human neutrophilic leukocyte granules. *J. Ultrastruct. Res.* 24, 343.

Dankmeijer, J. (1934). *De beteekenis van vingerafdrukken voor het anthropologisch onderzoek*. Thesis. Utrecht.

De Carli, L., Maio, J. J., Nuzzo, F. and Benerecetti, A. S. (1964). Cytogenic studies with alkaline phosphatase in human heteroploid cells. *Symposia on quantitative biology* XXIX, 223.

De Carli, L., Nuzzo, F., Santachiara-Benerecetti, A. S. and Morrow, J. (1968). Karyotypes of human cultured cell clones and their alkaline phosphatase activity. *Ann. N. Y. Acad. Sci.* 155, 1003.

Donaldson, D. D. (1961). The significance of spotting of the iris in mongoloids. *Arch. Ophthalm.* (Chicago), 65, 26.

Dooren, L. J. (1967). *Groei en geslachtelijke rijping bij cerebraal defecte kinderen. Growth and sexual maturation in cerebral defect*. Thesis, Leiden.

Down, J. L. H. (1866). Observations on an ethnic classification of idiots. *London Hospital Clin. Lect. and Rep.* 3, 259.

Edgren, J., Chapelle, A. de la and Kääriäinen, R. (1966). Cytogenetic study of 73 patients with Down's syndrome. *J. Ment. Defic. Res.* 10, 47.

Engler, M. (1949). *Mongolism (peristatic amentia)*. John Wright & sons Ltd. Bristol. Simpkin Marshall Ltd. London.

Ferrier, S. (1964). Enfant mongolien, parent mosaïque. *J. Génét. Hum.* 13, 315.

Finley, W. H., Finley, S. C., Rosecrans, C. J. and Tucker, C. C. (1966). Normal/21-trisomy mosaicism. Report of four cases and review of the subject. *Amer. J. Dis. Child.* 112, 444.

Fitzgerald, P. H. and Lycette, R. R. (1961). Mosaicism involving the autosome associated with mongolism. *Lancet* 2, 212.

Ford, C. E. (1969). Mosaics and chimaeras. *Brit. Med. Bull.* 25, 104.

Ford Walker, N. (1958). The use of dermal configurations in the diagnosis of mongolism. *Pediat. Clin. N. Amer.* May, 531.

Galbraith, P. R. and Valberg, L. S. (1966). Granulopoiesis in Down's syndrome. *Pediatrics* 37, 108.

Gall, J., Garn, S. M., Harper, M. and Stimson, C. W. (1970). Non-random chromosome losses in Down's syndrome. *Nature* 227, 499.

Garen, A. and Levinthal, C. (1960). A fine structure genetic and chemical study of the enzyme alkaline phosphatase of E. coli. I. Purification and characterization of alkaline phosphatase. *Biochem. Biophys. Acta* 38, 470.

Gelderen, H. H. van, Posthuma, J. H. and Haas, J. H. de (1954). Geboortegewicht en praematuritas in Nederland. *T. Soc. Geneesk.* 32, 443.

Gelderen, H. H. van, Gaillard, J. L. J. and Schaberg, A. (1967). Trisomy G/normal mosaics in non-mongoloid mentally deficient children. *Acta Paediat. Scand.* 56, 517.

Gelderen, H. H. van. (1967). *Personal communication*.

Giraud, P., Bernard, R., Stahl, A., Giraud, F., Hartung, M. and Lebeuf, M. (1963). Mosaïque chromosomique chez une mongolienne avec un Q.Iè à 0,85. *Pédiatrie* 18, 753.

Giraud, P., Bernard, R., Stahl, A., Giraud, F. and Hartung, M. (1965). Les mosaïques chromosomiques. A propos de quatre observations personnelles. *Pédiatrie* 20, 125.

Giroux, J. and Miller, J. R. (1967). Dermatoglyphics of the broad thumb and great toe syndrome. *Amer. J. Dis. Child.* 113, 207.

Greyerz-Gloor, R. D. von, Mauer, P. auf der, and Bergemann, E. (1969). Zytogenetische und phänomenologische Untersuchungen an 272 mongoloiden des Kantons Bern. *Schweiz. Med. Wschr.* 99, 1151.

Gustavson, K. H. and Ek, J. I. (1961). Triple stem line mosaicism in mongolism. *Lancet* 2, 319.

Gustavson, K. H. (1964). *Down's syndrome: a clinical and cytogenetical investigation*. Almquist and Wiksell, Uppsala.

Hall, B (1964). Mongolism in newborns. A clinical and cytogenetic study. *Acta paediat.* suppl. 154, 1.

Hamerton, J. L., Giannelli, F. and Polani, P. E. (1965). Cytogenetics of Down's syndrome (mongolism). I. Data on a consecutive series of patients referred for genetic counselling and diagnosis. *Cytogenetics* 4, 171.

Hayashi, T. (1963). Karyotypic analysis of 83 cases of Down's syndrome in Harris County, Texas. *Tex. Rep. Biol. Med.* 31, 28.

Hayashi, T., Hsu, T. C. and Chao, D. (1962). A case of mosaicism in mongolism. *Lancet* I, 218.

Hilgenreiner, H. (1925). Zur Frühdiagnose und Frühbehandlung der angeborenen Hüftgelenkverrenkung. *Mediz. Klin.* 21, 1425.

Hirsch, W., Leichsenring, G. and Lüers, Th. (1967). Klinische und Hautleistenbefunde bei Down-Syndrom mit Chromosomenmosaik (Mosaik-Mongolismus). *Mschr. Kinderheilk.* 115, 516.

Hsia, D. Y. Y., Inouye, T., Wong, P. and South, A. (1964). Studies on galactose oxidation in Down's syndrome. *New Engl. J. Med.* 270, 1085.

Huang, S. W., Emanuel, I., Lo, J., Liao, S. K. and Hsu, C. C. (1967). A cytogenetic study of 77 chinese children with Down's syndrome. *J. Ment. Defic. Res.* 11, 147.

Kaplow, L. S. (1955). A histochemical procedure for localizing and evaluating leucocyte alkaline phosphatase activity in smears of blood and marrow. *Blood* 10, 1023.

Kaplow, L. S. (1968). Leukocyte alkaline phosphatase cytochemistry: applications and methods. *Ann. N. Y. Acad. Sci.* 155, 911.

Kaufmann, H. J. and Taillard, W. F. (1961). Pelvic abnormalities in mongols. *Brit. Med. J.* 1, 948.

Kelly, S., Copeland, W. and Almy, R. (1967). Galactose -1-phosphate uridyl transferase in mongols. *N. Y. State J. Med.* 2714.

Koch, R., Share, J., Webb, A. and Graliker, B. V. (1963). The predictability of Gesell developmental scales in Mongolism. *J. Pediat.* 62, 93.

Kohn, G., Taysi, K., Atkins, T. E. and Mellman, W. J. (1970). Mosaic mongolism. I. Clinical correlations. *J. Pediat.* 76, 874.

Krivit, W. and Good, R. A. (1956). The simultaneous occurrence of leukemia and mongolism. Report of four cases. *Amer. J. Dis. Child.* 91, 218.

Lejeune, J., Gautier, M. and Turpin, R. (1959). Etudes des chromosomes soma-tiques de neuf enfants mongoliens. *C. R. Acad. Sci.* (Paris) 248, 1721.

Levinson, A., Friedman, A. and Stamps, F. (1955). Variability of mongolism. *Pediatrics* 16, 43.

Lindsten, J., Alvin, A., Gustavson, K. H. and Fraccaro, M. (1962). Chromosomal mosaic in a girl with some features of mongolism. *Cytogenetics* 1, 20.

Lowe, R. F. (1949). The eyes in mongolism. *Brit. J. Ophthal.* 33, 131.

Marks, J. F., Wiggins, K. M. and Spector, B. J. (1967). Trisomy 21 - trisomy 18 mosaicism in a boy with clinical Down's syndrome. *J. Pediat.* 71, 126.

Mauer, I. and Noe, O. (1964). Triple stem-line chromosomal mosaicism in Down's syndrome (mongolism). *Lancet* 1, 666.

Mellman, W. J., Oski, F. A., Tedesco, T. A., Maciera-Coelho, A. and Harris, H. (1964). Leucocyte enzymes in Down's syndrome. *Lancet* 2, 674.

Mellman, W. J., Raab, S. O., Oski, F. A. and Tedesco, T. A. (1968). Abnormal leukokinetics in 21 trisomy. *Ann. N. Y. Acad. Sci.* 155, 1020.

Mikkelsen, M. (1967). Down's syndrome at young maternal age: cytogenetical and genealogical study of eighty-one families. *Ann. Human Genet.* 31, 51.

Morgan, J. F., Morton, H. J. and Parker, R. C. (1950). Nutrition of animal cells in tissue culture. I. Initial studies on a synthetic medium. *Proc. Soc. Exper. Biol. Med.* 73, 1.

Mosier, H. D., Grossman, H. J. and Dingman, H. F. (1965). Physical growth in mental defectives. *Pediatrics* 36, 465.

Nadler, H. L., Inoùye, T. and Hsia, D. Y.-Y. (1967). Enzymes in cultivated human fibroblasts derived from patients with autosomal trisomy syndromes. *Amer. J. Human Genet.* 19, 94.

Nellhaus, G. (1968). Head circumference from birth to 18 years. *Pediatrics* 41, 106.

Neurath, P., DeRemer, K., Bell, B., Jarvik, L. and Kato, T. (1970). Chromosome loss compared with chromosome size, age and sex of subjects. *Nature* 225, 281.

Nichols, W. W., Coriell, L. L., Fabrizio, D. P. A., Bishop, H. C. and Boggs, Th. R. (1962). Mongolism with mosaic chromosome pattern. *J. Pediat.* 60, 69.

Nicolis, F. B. and Sacchetti, G. (1963). A nomogram for the X-ray evaluation of some morphological anomalies of the pelvis in the diagnosis of mongolism. *Pediatrics* 32, 1074.

Nowell, P. C. and Hungerford, D. A. (1960). Chromosome studies on normal and leukemic human leukocytes. *J. Nat. Cancer Inst.* 25, 85.

Øster, J. (1953). *Mongolism.* Copenhagen, Danish Science Press Ltd.

Pearson, H. A. (1967). Studies of granulopoiesis and granulocyte kinetics in Down's syndrome. *Pediatrics* 40, 92.

Penrose, L. S. (1954). The distal triradius t on the hands of parents and sibs of mongol imbeciles. *Ann. Human Genet.* 19, 10.

Penrose, L. S. (1963). Measurements of likeness in relatives of trisomics. *Ann. Human Genet.* (London) 27, 183.

Penrose, L. S. (1964). Unpublished observations, quoted in Penrose and Smith (1966).

Penrose, L. S. (1965). Dermatoglyphics in mosaic mongolism and allied conditions. *Genetics Today* 3, 973. Oxford: Pergamon Press.

Penrose, L. S. and Smith, G. F. (1966). *Down's Anomaly*. J. and A. Churchill Ltd. London.

Petit, P. and Gallez, A. (1969). Etude clinique et cytogénétique longitudinale d'un cas de mosaïque 46, XX/47, XX, G + (mosaïque de trisomie 21). *Helv. Paediat. Acta* 24, 582.

Pfeiffer, R. A. (1966). Résultats d'une étude cytogénétique et clinique de 312 mongoliens. Signification des translocations et mosaïques. *Ann. Génét.* 9, 94.

Raab, S. O., Mellman, W. J., Oski, F. A. and Baker, D. (1966). Abnormal leukocyte kinetics: an explanation for the enzyme abnormalities observed in trisomy-21 (Down's syndrome). *J. Pediat.* 69, 952.

Reinwein, H., Wolf, U. and Ising, H. J. (1966). Bericht über 3 Mosaikfälle mit G1-Trisomie (Mongolismus). *Helv. Paediat. Acta* 21, 300.

Richards, B. W., Stewart, A., Sylvester, P. E. and Jasiewicz, V. (1965). Cytogenetic survey of 225 patients diagnosed clinically as mongols. *J. Ment. Defic. Res.* 9, 245.

Richards, B. W. (1969). Mosaic mongolism. *J. Ment. Defic. Res.* 13, 66.

Ridler, M. A. C., Shapiro, A., Delhanty, J. D. A. and Smith, G. F. (1965). A mosaic mongol with normal leucocyte chromosomes. *Brit. J. Psychiat.* 111, 183.

Robinson, G. C., Miller, J. R., Cook, E. G. and Tischler, B. (1966). Broad thumbs and toes and mental retardation. Unusual dermatoglyphic observations in two individuals. *Amer. J. Dis. Child.* 111, 287.

Rosecrans, C. J. (1968). The relationship of normal/21-trisomy mosaicism and intellectual development. *Amer. J. Ment. Defic.* 72, 562.

Rosen, R. B. and Nishiyama, H. (1968). Leukocyte alkaline phosphatase in chronic granulocytic leukemia of childhood. *Ann. N. Y. Acad. Sci.* 155, 992.

Rosner, F. and Ong, B. H. (1967). Dermatoglyphic patterns in trisomic and translocation Down's syndrome (mongolism). *Amer. J. Med. Sci.* 253, 556.

Rubinstein, J. H. and Taybi, H. (1963). Broad thumbs and toes and facial abnormalities. A possible mental retardation syndrome. *Amer. J. Dis. Child.* 105, 588.

Schultze-Jena, B. S. (1959). Röntgenologische Merkmale des Beckens bei Mongolismus im Säuglingsalter, *Kinderärztl. Praxis* 27, 141.

Sergovich, F. R., Soltan, H. C. and Carr, D. H. (1964). Twelve unrelated translocation mongols: cytogenetic, genetic and parental age data. *Cytogenetics* 3, 34.

Shapiro, B. L., Gorlin, R. J., Redman, R. S. and Bruhl, H. H. (1967). The palate and Down's syndrome. *New Engl. J. Med.* 276, 1460.

Smith, D. W., Therman, E. M., Patau, K. A. and Inhorn, S. L. (1962). Mosaicism in mother of two mongoloids. *Amer. J. Dis. Child.* 104, 534.

Soltan, H. C., Wiens, R. G. and Sergovich, F. R. (1964). Genetic studies and chromosal analyses in families with mongolism (Down's syndrome) in more than one member. *Acta Genet. (Basel)* 14, 251.

Spicer, S. S., Horn, R. G. and Wetzel, B. K. (1968). Ultrastructural and cytochemical characteristics of leukocytes in various stages of development. *Biochem. Pharmacol. Special Suppl.* 143.

Swaak, A. J. (1967). Een onderzoek bij 125 kinderen met het syndroom van Langdon Down (mongoloïdisme). I. Ouders, zwangerschap, geboorte. II. Groei en ontwikkeling op de kleuterleeftijd. *Ned. T. Geneesk.* 111, 65 (I), 110 (II).

Taybi, H. and Kane, P. (1968). Small acetabular and iliac angles and associated diseases. *Radiol. Clin. N. Amer.* 6, 215.

Taylor, A. I. (1968). Cell selection in vivo in normal/G trisomic mosaics. *Nature* 219, 1028.

Taylor, A. I. (1970). Further observations of cell selection in vivo in normal/G trisomic mosaics. *Nature* 227, 163.

Taysi, K., Kohn, G. and Mellman, W. J. (1970). Mosaic mongolism II. Cytogenetic studies. *J. Pediat.* 76, 880.

Teplitz, R. L., Rosen., R. B. and Teplitz, M. R. (1964). Granulocytic leukemia, Ph-chromosome, and leukocyte alkaline phosphatase. *Lancet* 2, 418.

Tonomura, A. and Takehiko, K. (1964). Triple chromosal mosaicism in a Japanese child with Down's syndrome. *Acta Genet. (Basel)* 14, 67.

Tonomura, A., Oishi, H., Matsunaga, E. and Kurita, T. (1966). Down's syndrome A cytogenetic and statistical survey of 127 Japanese patients. *Jap. J. Human Genet.* 11, 1.

Trubowitz, S., Kirman, D. and Masek, B. (1962). Leukocyte alkaline phosphatase in mongolism. *Lancet* 2, 486.

Tsuboi, T. and Inouye, E. (1968). Chromosal mosaicism in two Japanese children with Down's syndrome. *J. Ment. Def. Res.* 12, 162.

Turpin, R. and Bernyer, C. (1947). De l'influence de l'héridité sur la formule d'Arneth. *Rev. d'Hémat.* 2, 189.

Valencia, J. I., Lozzio, C. B. de, and Coriat, L. F. de (1963). Heterosomic mosaicism in a mongoloid child. *Lancent* 2, 488.

Valentine, W. N., Follette, J. H., Solomon, D. H. and Reynolds, J. (1957). The relationship of leukocyte alkaline phosphatase to 'stress', to ACTH, and to adrenal 17-OH-corticosteroids. *J. Lab. Clin. Med.* 49, 723.

Valentine, W. N. and Beck, W. S. (1951). Biochemical studies on leucocytes. I. Phosphatase activity in health, leucocytosis and myelocytic leucemia. *J. Lab. Clin. Med.* 38, 39.

Verresen, H., Berghe, H. van den, and Creemers, J. (1964). Mosaic trisomy in phenotypically normal mother of mongol. *Lancet 1*, 526.

Walker, F. A. and Ising, R. (1969). Mosaic Down's syndrome in a father and daughter. *Lancet* 1, 374.

Warkany, J., Weinstein, D., Soukup, S. W., Rubinstein, J. H. and Curless, M. C. 1964). Chromosome analyses in a children's hospital. Selection of patients and results of studies. *Pediatrics* 33, 290 and 454.

Waxman, S. H. and Arakaki, D. T. (1966). Familial mongolism by a G/G mosaic carrier. *J. Pediat.* 69, 274.

Weinstein, E. D. and Warkany, J. (1963). Maternal mosaicism and Down's syndrome (mongolism). *J. Pediat.* 63, 599.

Weiss, L. and Wolf, C. B. (1968). Familial C/G translocation causing mitotic nondisjunction. *Amer. J. Dis. Child.* 116, 609.

Wetzel, B. K., Spicer, S. S. and Horn, R. G. (1967). Fine structural localization of acid and alkaline phosphatase in cells of rabbit blood and bone marrow. *J. Histochem. Cytochem.* 15, 311.

Wilkins, L. (1965). *The diagnosis and treatment of endocrine disorders in childhood and adolescence.* Thomas, Springfield, 3rd. ed.

Wijn, J. F. de and Haas, J. H. de (1960). Groeidiagrammen van 1-25 jarigen in Nederland. *Verhandeling Ned. Instit. Praev. Geneesk.* XLIX, Leiden.

Youden, W. J. (1951). *Statistical methods for chemists.* John Wiley and sons, Inc., New York, p. 41.

Yu-Feng Hsu, Schweger, A. J., Nemhauser, I. and Sobel, E. H. (1965). A case of double autosomal trisomy with mosaicism: 48 XX (trisomy 18 + 21). *J. Pediat.* 66, 1055.

Zapella, M. and Cowie, V. (1962). A note on time of diagnosis in mongolism. *J. Ment. Defic. Res.* 6, 82.

Zellweger, H. and Abbo, G. (1963). Chromosal mosaicism and mongolism. *Lancet* 1, 827.

Zellweger, H., Abbo, G., Kay Nielsen, M. and Wallwork, K. (1966). Mosaic mongolism with 'normal' chromosomal complement in white blood cells. *Humangenetik* 4, 323.